AN EAGLE
NAMED FREEDOM

AN EAGLE
NAMED
FREEDOM

My True Story of a Remarkable Friendship

JEFF GUIDRY

wm

WILLIAM MORROW
An Imprint of HarperCollinsPublishers

I have changed the names of some individuals in order to preserve their anonymity, and in some cases have compressed timelines in order to maintain narrative flow. The goal in all cases was to protect people's privacy without damaging the integrity of the story.

HarperCollins books may be purchased for educational, business, or sales promotional use. For information please write: Special Markets Department, HarperCollins Publishers, 10 East 53rd Street, New York, NY 10022.

FIRST EDITION

Library of Congress Cataloging-in-Publication Data
Guidry, Jeff.
    An eagle named freedom : my true story of a remarkable friendship / Jeff Guidry.
        p. cm.
    ISBN 978-0-06-182674-0
    1. Bald eagle—Washington (State)—Biography. 2. Guidry, Jeff. 3. Human-animal relationships—Washington (State). 4. Wildlife rehabilitation—Washington (State). 5. Cancer—Patients—Washington (State). I. Title.
    QL696.F32G85 2009
    639.9'78943092—dc22

                                                        2009031600

10 11 12 13 14   OV/RRD   10 9 8 7 6 5 4 3 2 1

In memory of Kaye and Bob
For Lynda
And for Dream Flyer

AN EAGLE
NAMED FREEDOM

INTRODUCTION

THE EAGLE WAS YOUNG—and she was badly injured. The bird was about three feet tall and probably weighed eight pounds, though the long plumage of a young eagle made her appear bigger. Her feathers and her eyes were dark brown.

It was August 12, 1998. The eagle looked up at me and my old life was over, a new second life begun. I had been volunteering at the wildlife center for almost two years, and I'd helped with hurt eagles before, but this time was different. Freedom would become my friend and my teacher. With her I would discover and deepen parts of myself. Two years later, when I was diagnosed with stage 3 non-Hodgkin's lymphoma, my third life began. But I'm getting ahead of myself.

❧ ❧ ❧

ON THE DAY SHE was brought in, two other people were working with the eagle, good friends of mine, the director of Sarvey Wildlife Care Center, Kaye Baxter, and Sarvey's most dedicated volunteer, Crazy Bob Jones. I couldn't tell by looking how badly the eagle was injured.

Kaye said, "Let's check her out." I picked up the eagle and held her as gently as I could. Kaye carefully stretched out one wing, then the other; she ran her fingers along the bones of the wing. "They're both broken," she said. "Jeff, can you take her to the vet?"

I looked at the eagle and said, "Yes." I was drawn to her more than any other wild creature I had seen. I didn't question why.

Even now after eleven years of a growing connection, sometimes the immediate power of my feelings for Freedom, as we came to call her, can seem weird to me. Feeling invested is a natural part of rescue and caring for wild ones, and I took Freedom's well-being personally right away. But there was an intensity that went beyond personal.

On the drive to the vet, I could see pain and trust in her eyes. I hoped she could sense my caring as I spoke softly to her. Our conversation went beyond the words, and our unspoken connection was the beginning of the miracle that continues to unfold—a miracle of mutual healing, a miracle of deep friendship.

The next phase of the miracle was a hard phase for me

and those I love. In the spring of 2000, I was diagnosed with stage 3 non-Hodgkin's lymphoma. There were eight months of grueling chemotherapy. Freedom came to my rescue.

As powerful as that time was, it didn't occur to me to write much about it. I wrote a few pieces for the Sarvey website, contributed to a few blogs, appeared on a radio show—and then on March 6, 2008, I wrote the bones of our story in an e-mail to my friend Gayle, who works in wildlife activism. She had asked me about Freedom, and when I told her that there was a story to go with the pictures I had sent her, she said she'd love to see it.

As I wrote her the story, I vividly remembered my first encounter with Freedom and how Freedom had been one of my most important allies during my battle with cancer. It felt good to share this information with a friend in the work. But I would never have expected the long-range consequences of that e-mail.

Hey Gayle, here is the information you asked for about Freedom and me. When Freedom came in she could not stand. Both wings were broken, her left wing in 4 places. She was emaciated and covered in lice. We here at the Sarvey Wildlife Care Center made the decision to give her a chance at life, so I took her to the vet's office.

From then on, I was always around her. We had

her in a huge dog carrier with the top off, and it was loaded up with shredded newspaper for her to lie in; I used to sit and talk to her, urging her to live, to fight, and she would lie there looking at me with those big brown eyes.

We had to tube feed her for 4–6 weeks, and by then she still couldn't stand. Finally the decision was made to euthanize her if she couldn't stand in a week. It looked like death was winning.

She was going to be put down on a Friday, and I was supposed to come in on that Thursday afternoon. I didn't want to go to the center that Thursday, because I couldn't bear the thought of her being euthanized; but I went anyway, and when I walked in everyone was grinning from ear to ear.

I went immediately back to her dowel cage. There she was, standing on her own, a big beautiful eagle. She was ready to live. I was just about in tears by then. That was a very good day.

We knew she could never fly, so the director asked me to glove train her. I got her used to the glove, and then to jesses (they are a kind of leather leash around each leg), and we started doing education programs for schools in western Washington. We wound up in the newspapers, radio (believe it or not) and some TV. Miracle Pets even did a show about us.

In the spring of 2000, I was diagnosed with non-Hodgkins lymphoma. I had stage 3, which is not good (one major organ plus everywhere), so I wound up doing 8 months of chemo. Lost the hair—the whole bit. I missed a lot of work. When I felt good enough, I would go to Sarvey and take Freedom out for walks. She would also come to me in my dreams and help me fight the cancer—time and time again.

Fast forward to November 2000, the day after Thanksgiving, I went in for my last checkup. I was told that if the cancer was not all gone after 8 rounds of chemo, then my last option was a stem cell transplant. Anyway, they did the tests; and I had to come back Monday for the results. I went in Monday, and I was told that all the cancer was gone. Yahoo!

So the first thing I did was get up to Sarvey and take the big girl out for a walk. It was misty and cold. I went to her flight and jessed her up, and we went out front to the top of the hill. I hadn't said a word to Freedom, but somehow she knew. She looked at me and wrapped both her wings around me to where I could feel them pressing in on my back (I was engulfed in eagle wings), and she touched my nose with her beak and stared into my eyes, and we just stood there like that for I don't know how long. That was a magic moment. We have

been soul mates ever since she came in. This is a
very special bird.

I never forget the honor I have of being so close
to such a magnificent spirit as Freedom's. Hope you
enjoy this.

Gayle told me she had been so moved by my e-mail that she wanted to forward it along with pictures. I said, "Sure," and promptly forgot about the whole thing.

The recipients of Gayle's forward were moved too. They passed the story on. In a couple of weeks, thousands of messages flooded my e-mail in-box from all over the world. People wrote because they had cancer, because they loved wild creatures, and because they just wanted to be near the miracle. So many messages came in that I had to open up a new e-mail account.

I knew then that it was time to put the story of an eagle and a man saving each other's lives out to the world. I never would have believed until that day in 1998 that there was an eagle on her way to meet me. And I never would have imagined how deeply she would change my life. I'd be content if everyone who reads this finds a little hope that miracles are real.

JEFF GUIDRY

CHAPTER ONE

ONE OCTOBER DAY IN 1992, I suddenly really saw an eagle for the first time. I was driving through the mountains when a huge shadow came over the car. I looked out and up. There was an eagle just ahead of me. It was flying so close to the car that I swear I could see every white feather in its tail gleaming in the early morning sun. I was transfixed.

After that first astonishing sighting, I began to see eagles everywhere in my new home in the Pacific Northwest. My partner, Lynda, and I had moved to eagle country in 1989. I was fascinated by eagles and started reading about the birds and the beliefs of the native people who hold them sacred.

Pretty soon I was a walking eagle encyclopedia, so I was intrigued in 1992 when Lynda told me she'd seen something about eagle watching on the news. The Eagle Watchers' mission is to educate the public about bald eagles. They also wanted to have the public view the birds in designated areas on the Skagit River, one of the largest gatherings of bald eagles in the lower forty-eight states, where the birds follow the salmon to the Skagit feeding grounds. It was important to have designated areas, because if people stop all along the eleven-mile stretch of the river between Concrete and Marblemount and disturb feeding eagles, the eagles will leave the area, which is prime winter feeding habitat for them.

Loving animals was something Lynda and I had always shared. We'd both had many pets before we met. Our first cat, Baby Boy, had sauntered in when we lived in California, where I was a professional musician for years until I burned out on it. Baby Boy loved to greet and hang out with the band when we'd come to my house around 3:00 A.M. after playing. He'd share our *machaca* burritos from the all-night Mexican food joint across the street, sitting on the couch with us just like one of the guys, getting some from each person. We thought about getting him a shirt that said IT'S OK, I'M WITH THE BAND.

Lynda would come out in the morning to get ready for work and find one content Baby Boy soundly asleep on the couch. When he came into the kitchen for his breakfast, she'd tell him she knew he'd had a hard night.

So, since we had always had a shared passion for our pets, eagle watching together was a natural for us.

In December 1992, Lynda and I stood in the rain and cold on the banks of the Skagit River about an hour north of our home—to watch and listen to the eagles. The river was gray, the light only a little less gray. Lynda and I were barely sheltered by a few magnificent cedars. A handful of tourists and Seattle locals stopped, stayed a minute or two, and left. Lynda and I held our ground.

For minutes at a time I almost couldn't move. The sound of the river and the quality of the wet gray air seemed to transport me to another realm. The workaday world was gone. The sounds of modern life were gone. I could have been standing on the shore of the Skagit centuries earlier. And despite the bone cold, I could have stood there for days.

I knew that the Skagit had been designated a Wild and Scenic River by the federal government since 1978 and was therefore protected from any development or use that would keep it from being a free-flowing river. I also had learned a few of the First Nations' beliefs about why it was considered sacred. But being still in that place made the magic come alive for me.

We volunteered for the Eagle Watchers for four years. Then in autumn 1995, Melanie Graham, the river ranger, handed me a pamphlet from the Sarvey Wildlife Care Center and said, "Check this out. You might enjoy this."

Lynda and I met the director of Sarvey, Kaye Baxter, at the Bald Eagle Festival in Concrete, Washington, a small-town festival held in the local high school gym in early February 1996. Lynda and I had gone up together to meet her.

Kaye and the eagle Yakala were in the gymnasium along with many other exhibitors. Yakala was in a six-by-six-foot chain-link enclosure. Kaye had just lowered the sheet over his pen so they both could take a break.

A man came up and started to lift the sheet to see. Kaye whirled around and snapped, "Put the damn sheet down, stand back, and show a little respect for the eagle!" He dropped it and disappeared into the crowd. Without missing a beat Kaye turned back to us, smiling as if nothing had happened. She began to explain what Sarvey is and does. *Cool*, I thought, *that's serious spunk.*

Kaye was a diminutive woman in her midfifties with wavy blond hair. She was five feet two or three inches and weighed about a hundred pounds, with the attitude of a warrior. I immediately liked her. We talked for a moment and I told her about Melanie Graham, the Skagit River ranger, sending me her way.

Kaye filled me in about Sarvey. She said the center was a nonprofit wildlife care center whose goals are rescue, rehab, and release. In some cases, permanent injury or habituation to humans can make animals unsuitable for release. She told me what kind of wild ones they treated and spoke about a few of the permanent residents, like her red-tailed hawk and soul mate, Mellow Yellow; a Patagonian cougar named Sasha; and a barn owl named Mum.

There was no formal orientation for new volunteers. "Dr. Judy runs Thursdays, and she'll show you around," Kaye assured me. Thursdays were great for me, since I had that day off.

I HAD NO IDEA what I was walking into that first day I set foot on the muddy drive leading up to Sarvey Wildlife Care Center. It was a bleak and gray Thursday in late February 1996, a few weeks after the festival. I took old, winding, two-lane country roads just at the edge of the foothills of the Cascade range, down into the lush Stillaguamish Valley. The sky was a crackled gray, and the weather was one of my favorite kinds—cold and a little damp. The thirty-five-minute drive gave me time to think. I wasn't totally sure that I wanted to get involved in wildlife rehab, but I knew that I owed it to myself to find out.

I saw the wildlife center perched on top of a hill surrounded by giant cedars and firs. There were two ponds at the bottom of the hill. A mallard took off from the dark water. A dirt drive led the way uphill to the center. I felt excited and a little nervous as I parked and walked up the drive. There was a beautiful old schoolhouse to my right and some drab buildings straight ahead and on the left.

I knocked on the front door of the schoolhouse. Kaye opened it. With scarcely a word, she waved me toward a building directly behind me with huge deer antlers over the entrance. I knew enough about the Sarvey values to know that the antlers had not come from a hunter, but would have been collected from the forest floor after the deer had shed them.

I opened the screen door and walked straight into a long kitchen in which three or four people were cutting up apples, bagging worms, and slicing meat on a long table. One of them, a bright-eyed, short-haired woman, looked up and said, "Hi, I'm Judy." Her grin put me at ease. I noticed right away that the group of volunteers was a mixed bag, that it crossed age, gender, and ethnic lines. I could sense dedication in their manner as they carefully prepared the food for their wildlife patients.

Judy (or Dr. Judy as the volunteers called her—though she hated it) showed me around. We went into the back room, which was filled with aquariums, most of them empty. She said they wouldn't be empty for long; the first babies of spring would begin to arrive in March. "That's the beginning of baby season," Judy said.

There was a big four-by-four-foot cage with wooden bars on the opposite wall. "This is the dowel cage. It's for eagles," Judy said. "The wooden bars don't damage their feathers. And they are easier to catch in there." There were six big green cages stacked one on one, lined up along one wall, most of them occupied. "Those," Judy said, "are known, strangely enough, as the large greens. They're for raptors; medium-sized mammals like beavers, porcupines, fawns, opossums; and ducks and swans. They're all occupied at the moment, even though you can't see some of the creatures because we keep sheets hanging down in front of the doors. When the injured come in, this is a very frightening place for them. They are wild animals, hurt, and in a strange place with people handling them. We try everything we can to keep these patients as calm as possible. The sheets block their visual field and help them feel safe."

I looked in at a saw-whet owl, and a raccoon. The big-eyed owl was beige and dark brown, and no more than seven inches tall. It stared steadily at me. The raccoon turned its back on me. I imagined that in a few weeks I would be allowed to actually work with these creatures. I was wrong.

Judy had a different idea. "Let's go get to work." She taught me how to clean the cages. Every cage was cleaned and sterilized every day, with fresh food, water, and bedding put in. There were no exceptions to this rule.

I was pleased not to have to wait to try out the work. I started my cleaning with the small-mammals aquariums. Judy pointed to an occupied aquarium on a shelf. "We'll start with this guy," she said. "Grab one of those clean aquariums behind you," Judy said. "Put in food, water, and fresh bedding. The food and water are in the kitchen. The bedding is in the storage room."

I followed orders. When the cleaned aquarium was all set up, Judy said, "Now, you get to reach in and grab the flying squirrel in that aquarium on the shelf. Just reach in and get him. . . ."

I'd already reached in and had him in my hand. Before I knew it, vicious tiny teeth had sunk into my finger. Blood flowed before I felt the pain. My finger looked as though a sewing machine had gone after it.

I put the squirrel quickly and carefully in the clean aquarium and slapped the cover on. Then I wiped a little blood on my jeans and got on with it.

That day I also helped with a great blue heron and went on a deer rescue with Megan, a longtime volunteer. We found the deer down in the woods behind the house of the woman who had called us. The doe was in bad shape, though we could see no obvious wounds or breaks. We loaded her into the truck and headed back to Sarvey. Megan was in back riding with the doe. About halfway to Sarvey I could hear tapping on the rear window. I pulled over. "She died," Megan said quietly. *This is the real deal*, I thought.

Later that afternoon, Judy came to me and said, "Now let's go for one of the big gals." She led me to an eagle in a large green and demonstrated how to take a blanket, kneel down, and try for the angle that would work, preferably with the bird's back to you. "Go in fast and cover the eagle and hold her down, all in one motion, then take hold of both legs. You must have control of the legs or you are screwed," she warned me. "You'll have to start over and that is no fun with an agitated eagle. Once you get the legs under control by gripping one leg in each hand, you move a wing under each arm; then you have total control.

"These birds are incredibly strong and fast; those talons are weapons that can inflict serious damage. Staff and volunteers have been sent to the hospital because of getting 'taloned.' If you handle raptors on a regular basis, you will get nailed. It just happens. As we say here, 'if it doesn't bleed, it doesn't count.'" I'd already learned that with the squirrel, so I was ready.

Judy stepped back and held the big bird while a young female volunteer efficiently cleaned the cage, lined the floor with newspapers, and set the cleaned perch back in. Then Judy set the bird gently back in the cage. "Getting the blanket off an eagle is sometimes almost as hard as putting it on the bird," Judy said. I was impressed with how smoothly they worked.

I was itching to try for myself, but Judy said, "I want to take you through the process so you have it fixed in your mind and mental muscles."

I waited.

"Take a deep breath," she said. "Now visualize how it's going to go with you capturing the eagle. See the whole process in your mind. Get your breathing steady and mind focused. Then when you're ready to make the catch, do not hesitate."

We walked over to another large green with a bald in it. I stopped for a minute and let my breathing slow, then I visualized the catch as it was supposed to go. Judy had said that if the capture didn't go as it should, I'd better have a backup plan.

My adrenaline was flowing. I held my focus and swooped in with the blanket to make the capture. As soon as I had the eagle in my arms, I felt a rush I'd never felt before. I held the bird, and I swear I could have been holding a bear—all muscle, instinct, and power. I was pumped—me and my first eagle.

We finished the cleaning routine, and I put the eagle back in its cage. "You might be a natural," Judy said. "We might keep you."

Just then, Kaye appeared in the doorway and said, "Have you met Sasha, the cougar?"

"Not yet," I said eagerly. I was still high from my first eagle.

"Go out back," she said. "I can't go with you because Sasha hates me."

"By the way," Kaye added as I walked out the door, "she likes grass."

I took a left and saw a spacious enclosure. A big tawny cougar with a gorgeous tail walked toward me. She moved like silk.

I strolled over, knelt down in front of her, and said "Hi, Sasha, how are you?" She rubbed her face on the chain link, and I could hear her small motor going, so I reached in between the links and scratched her cheeks and chin.

Sasha's fur was short, coarse, and thick. She had a distinct odor that was kind of sweet, like nothing I had ever smelled before. Her motor got louder the more I scratched her. Imagine your average house cat purring and amp it up ten times, then make it deep and resonant. I picked handfuls of grass and fed them to Sasha a few blades at a time. She nibbled delicately at the grass. Watching her take the grass so gently from my hand, I studied her large incisors. I thought, *Damn, those are some impressive teeth.* I was fascinated by her elegant power.

CHAPTER TWO

On my next visit a week later, Kaye Baxter invited me into her home, the converted old schoolhouse at Sarvey, for coffee.

"I want to tell you how this all happened," she said.

"In 1980, when I was forty, I lived in Everett, Washington. A couple of the neighborhood kids found an injured saw-whet owl and brought it to me. I'd worked with Joni Butler at the Seattle Wild Bird Clinic, so I knew the basics of taking care of the little owl. Those kids, the owl, and I were the beginning of Sarvey.

"Soon, people began to hear about what I was doing and brought me other hurt raptors. The neighborhood kids continued to bring in baby opossums, raccoons, and squirrels. I had to find bigger quarters for the creatures."

Kaye told me that in 1987 she bought five acres just south of Arlington, Washington, at the top of a low wooded slope, to give thousands of what she called "our brother and sisters" or "the wild ones" a chance. The center had gotten its name in 1988 in memory of Kaye's friend Bill Sarvey, a Washington Department of Wildlife agent. "I've kept my promise to the wild ones," she said.

Within a few weeks, I had decided to make my own promise to Sarvey's creatures. The clinic asks for four hours a week, and that was how I started, every Thursday. It was the new challenge I'd been looking for. I'd be making an immediate difference in the lives of wild ones, and the work would teach me things I'd never learn in books or on the Internet. I would be part of work that mattered.

Over my first months as a volunteer I learned that Sarvey provides not just rescue, rehabilitation, and release—and shelter when release isn't possible for wild ones. The center also gives merciful death to those animals who are beyond help, and some raptors who can't be released become part of educational presentations.

Kaye's goal and ours was to keep the wild ones safe and reproducing, thereby helping to balance the ecology in our part of the Pacific Northwest—in Snohomish, King, Skagit, and Pierce counties. Kaye told me more than once—with her customary passion—that as human development had taken over more and more of the area, mating, breeding, and hunting grounds for the wild ones had shrunk exponentially. I knew that habitat destruction was the eagles' greatest threat and that we are their only natural predators. DDT, which causes the eagles' egg shells to soften, almost wiped out the bald eagle until the pesticide was banned in 1972; the only reason the species survived was because the mating pairs lived long enough to outlast the poison. Kaye always said that more people was the problem—that as the human population had grown, the damage caused by vehicles, carelessness, and pure malice had escalated.

As I continued to volunteer at Sarvey over the next few years, I learned more about Kaye. She was more than the director of Sarvey. She was the fire, fuel, queen bee, and servant of the center. Sarvey was her life. I would learn from her, respect and love her, and fight fiercely with her. We were two strong-willed people with a passion for those who cannot speak out for themselves in human tongue.

As Kaye told me more about her initial motivation for wildlife rescue, I began to understand a deep aspect of her spirit. Her grandmother was Cherokee, and Kaye carried with her a belief in native ways that she taught to many of us. For example, the Northwest native peoples regard the eagle as a messenger of peace, and many Northwest tribes scatter eagle down as a symbol of friendship and good intentions in front of guests at welcoming ceremonies.

Kaye had been diagnosed with breast cancer in the late 1970s. Out of all the women in her treatment group, she was the only survivor. The radiation treatments were brutally harsh. The doctors had administered the radiation from the front. It penetrated all the way through to her back and left black peeling chunks of skin as though she had been burned. Her lungs and throat were scarred, and she lost one breast, but she had her life. She knew she had to give back.

Kaye had come from big money. In fact, she often referred to her former self as Mrs. R.B.—Mrs. Rich Bitch. She said that cancer put everything into perspective. An injured saw-whet owl gave her the focus for her payback. Her gifts as a teacher and a performer would help her gratitude become action.

Kaye's programs aimed at education and prevention. She wanted to end human damage to wildlife—to eliminate habitat destruction, carelessness, ignorance, and cruelty. She believed in introducing humans to their wild brothers and sisters.

All this required more than vision and money. The veterinarians donate their time. Sarvey's corporate officers and board of directors take no salaries. On the front lines, more than one hundred volunteers keep Sarvey's doors open. The spring and summer are our busiest times. Every cage or enclosure gets cleaned every day, and all animals get fresh water and food every day. Some volunteers clean and feed; others work with the staff handling animals and administering medications. Some concentrate on maintaining the buildings.

As I worked with these dedicated people, I saw a young

woman with black-and-pink Goth hair and multiple piercings coming back twice a week to clean out smelly cages without complaining. I worked alongside high school kids who cared more about the animals whose food they were preparing than their own social lives. I watched an old man nearly crippled by arthritis come every Thursday and clean cages. He could barely get back up after kneeling on the floor.

Volunteers at Sarvey need more than love for the wild ones. We have to have courage. Scared and hurt animals lash out in fear, and they are powerful and fast. I have had more than a few experiences with beaks, claws, teeth, and talons. But the payoff is spectacular—especially the times I get to make a release.

When I release an eagle or watch an opossum scamper back into the woods or a squirrel shinny up a tree, I know I've made a difference in the life of one wild creature. I can almost feel what they feel—the rush, the freedom, the home-coming.

One day that first year, after I had been coming to Sarvey for about six months, a staff member put up a bulletin board in the staff lounge. The lounge is a beat-up room with a table, chairs, and random equipment stashed in the corners.

I took my break and looked up at the bulletin board. I could find myself in so many of the responses to the statement:

I VOLUNTEER BECAUSE . . .

I leave the crowded city to come and be with the people and animals that I connect so deeply with. It is an incredible feeling and impossible to explain. This is my sanctuary. —A.

Animals don't judge.

Da $ +
Da Fame +
Da Respect from Peers Heck, yes

I can talk to the birds and animals with nobody laughing. p.s. to meet cool people. —C.

I get to do stuff you can't do anywhere else. How many people have held an eagle in their arms or cuddled an opossum. —J.

I volunteer at Sarvey because I love animals more than anything I have in my entire life. I also want to gain skills that I can use in my career working with animals. Plus working at Sarvey is a blast and a half. —Cody

To meet chicks. —M.

Because of my love for nature. —D.

Cool stains on my pants . . . to get dirty! —H.

I love Sarvey! I come here to be around animals that I love, meet new people, get away from my normal life and mostly *have something that is my own.* How many people can say they got to hold an owl today?! —A.

I first met Sarvey's most dedicated volunteer, Crazy Bob Jones, within the first week of being there. He was rail-thin, with ratty hair, long nails, nasty teeth, and a grimy baseball hat. Judy told me he'd gotten his name when he climbed a steeply pitched roof in a howling lightning-and-wind storm to rescue a red-tailed hawk.

His beady eyes looked intently into mine. We both knew right away we were kindred spirits. When I found out he lived on a diet of Marlboro 100s (red pack), M&Ms, McDonald's hamburgers, and massive amounts of coffee—not necessarily in that order—I knew he was made of tougher stuff than most humans.

Bob was smart, funny, and fierce, and I caught on right away that he knew ways to be with the animals that came out of lots of experience and his passion for his patients. I once watched him reach into a cage and gently take a newborn raccoon from its severely injured mom. He held the baby in his palms, and I swear the look on his face was purely maternal.

That deep love could go feral in a heartbeat. He was a ferocious defender of the wild ones—especially raccoons. Crazy Bob would go where only a true animal would go. And that is because he was a raccoon; that was his totem. He was crafty, smart, persistent—and he ate all his food with his fingers, just as raccoons do.

That first year, we went on dozens of rescue operations. He taught me almost everything he knew—what he couldn't teach me was to do what any sane person wouldn't have. Once we were called out to capture a river otter believed to be under someone's house. When we got to the house, Bob grabbed his rescue net and flashlight from the fully equipped ambulance (our rescue vehicle). Without hesitation, he wormed his way into the crawl space under the house. His skinny body could barely fit, but that didn't stop him. Neither did the knowledge that a trapped and furious river otter with razor-sharp teeth was the equivalent of a living chain saw.

Luckily for Bob, there was no otter in sight. Despite his personal and highly effective technique for netting wildlife, he would have been jammed tight, with no place to turn around and face-to-face with an animal who, however it appears in cutesy photos, is anything but cute when threatened.

The Seattle police loved Bob. Anytime they had trapped or hurt wildlife, they'd call him. Bob once rescued a coyote from a downtown Seattle office building. It had wandered in and become scared and was hiding out in an elevator.

A lot of media people were present. They too loved Bob

and he reveled in the attention. He waltzed right in, knelt down, and softly spoke to the coyote. Without a fuss on the 'yote's part, Bob put him in the carrier and took him from the dangers of so-called civilization. That was the lead story on all the evening news stations that night.

I learned fast, especially about how naïve most humans are about wild ones. The following spring, we got countless calls because of raccoon families in people's attics. The phone calls always sounded the same: Bob, after listening a few minutes, would patiently explain that the mother raccoon was simply going to raise the babies for a month or so, then leave: "All you have to do is wait the mother raccoon out." Then, just about every single time, he'd hang up and say, "They didn't listen. Let's go."

We would have preferred to spare any wild one the stress of human interference, but we knew it was better for us to remove the raccoon family gently than let the frantic human take more drastic measures. Our equipment included welding gloves in addition to the net and carriers, because though we were committed to doing the removal gently and with the least disruption, the fiercely protective mother raccoons didn't know that we were friends.

I learned something on every ambulance run. One time that spring, we encountered a mother raccoon who'd been hit by a car. Her back legs were broken. She had been dragging herself around taking care of her babies—so much that all the fur and skin had been scraped down to the bone. Bob looked down at her and said, "You've got to be one of the finest mothers on the planet."

Later that night, I remembered something I'd read: "The Aztecs called the female Raccoon 'see-oh-at-la-ma-kas-kay,' which means 'she who talks with gods,' and the female Raccoon with cubs was called 'ee-yah-mah-tohn,' which translates to 'little old one who knows things.'" I imagined that Bob was a male version of "ee-yah-mah-tohn."

He was one of the most intelligent men I'd ever met and one of the clean angriest about mistreatment of animals and the earth—and easily the sneakiest. It would be years before any of us would find out all he had done for Sarvey.

CHAPTER THREE

SARVEY WAS CHANGING MY LIFE, and those changes came home with me when, only months after I started, Bob gave me and Lynda one of the biggest gifts we've shared. He had rescued a tiny squirrel on an ambulance run to Bellingham. He brought the squirrel to Sarvey and called me to come see it. It was a tiny fellow with large teeth. We put the squirrel in with a gang of other baby squirrels and when they were old enough to be released, Lynda asked me if we could release them at our house and Kaye gave permission.

Lynda and I took the carrier to a patch of trees and grass that ran for a mile down past our house. We set the carrier on the grass and opened it. All the squirrels scooted out except the little runt with the vicious teeth. He sat in the carrier—for hours. Finally, we gently shook the carrier. He summoned enough squirrel chutzpah to venture out and scramble up a cedar—but only to the first branch. That was enough for him.

We checked on him again before dark. He hadn't moved. I felt terrible leaving him out there, but I figured, hey, he was a squirrel. He'd have to figure it out.

First light, Lynda and I looked for our new squatter. He was in the same spot, eyes staring straight ahead. He didn't move. I thought I knew what he was thinking: don't move, something awful might see me. *Good grief,* I thought, *he's terrified.* But I had to go to work, so I wished him luck and headed out.

When Lynda came out to go to her job an hour later, the little squirrel was hanging by his nails from the carport roof. She gently pushed him back up on the roof and got in the car. When she looked back at the garage, she saw the squirrel sitting absolutely motionless on the roof.

She got home from work before I did. Tragedy had struck. The squirrel lay sprawled under the big fir tree, on the grass. His little eyes were closed. He didn't seem to be breathing. She bent to check on him. He sprang onto her arm and raced up to hide in her long hair. She could feel him trembling against her neck. *That's enough of the wild for this guy,* she thought.

We had a squirrel. Better said, he had us. I checked with Kaye, and she said it was OK for him to rehab with us.

He needed a name. Lynda said, "He's so timid, let's call

him Timmers." That didn't last long. Timmers became Mr. Timms, and that stuck, though sometimes when he was his bad boy self, we called him Tiny Tooth, or the Terrible Tooth.

If ever two people were willing to take their place in the food chain, it was Lynda and me in relation to that squirrel. Mr. Timms was my at-home connection to the wild ones. And he certainly wasn't tame. If it could be chewed, Mr. Timms would chew it. If it could be pooped on, he'd poop on it. We built him a walk-in condo with two houses, two water stations, and a litter box that had oatmeal in it. Oatmeal works great. It clumps when it gets peed on; just grab a clump and you're done. Too bad we didn't think to put it all over the house.

However afraid Mr. Timms had been about going off on his own, he was never particularly timid with us. He'd rattle his condo door at about seven every morning—"Hey, wake up. It's party time!"

I'd open his door, and Mr. Timms would run up my arm and sit on my shoulder. Off we'd go into the living room, him bouncing up and down—boingy boingy boingy, his little head bobbing. He was ready to get on the squirrel highway and motor, to take apart coffee tables with his teeth, to pee and poop on window ledges. Mayhem in the morning was his motto.

The squirrel highway was our two chairs, a couch, and three barstools. We hadn't realized till Mr. Timms took over that we had a squirrel racetrack in our living room. His most spectacular cruising activity was launching himself off my arm onto the top of a barstool where he would attach to the edge of the seat and race in dizzying circles while the barstool rocked from side to side.

Mr. Timms revealed his spiritual side at Christmas that year. First of all, he respected the Christmas tree and never climbed it. And then, once Lynda had put out her Nativity scene, Mr. Timms kidnapped the baby Jesus out of the manger and stashed it in his house. We always called it the Moosetivity scene because it was rich with nonhuman animals, but Timms never had a particular interest in anything but the baby human. Lynda put the baby Jesus back in the manger. Mr. Timms snatched the baby Jesus out with his teeth and carried him down the hall. It happened again and again. Finally Lynda gave up and bought a new baby Jesus. Since he already had one, he let hers alone.

As I continued my work at Sarvey, my relationship with the cougar, Sasha, deepened. She was a Patagonian cougar who was purchased as a kitten from a dealer in Texas. "It was to fulfill some guy's dream of being a big man," Kaye said. "It happens all the time to exotic birds, mammals, and reptiles. It happens when an asshole thinks his money gives him the right to rob the freedom of one who belongs in the wild."

When we human beings forget our place, we doom a captured wild one to a life behind bars. A cougar needs at least between one hundred and four hundred square miles to live. Both male and female need the room to hunt, breed, and run free. The female needs room to give birth and raise and teach her young. Sasha's enclosure was still a jail cell, but we could give her some space to roam, and plenty of tree trunks and logs to hide behind. The one thing we couldn't give her was release, as she was declawed and habituated to humans—and not native to Washington. That old man in Texas had destroyed the lives of tigers, lions, bears, and other cougars, and when I thought about Sasha's life being sold for a few bucks and the freedom she never knew taken away, I wanted to beat the crap out of that guy. I wanted to watch him die.

At least Sasha was with people who respected her true nature, and as I got to know her I learned more about her kind. The tawny cougar is known as the cat with many names. Mountain lion. Puma. Catamount. Firecat. Cat of the Gods. There are eighty-three names that we know of—some from the indigenous peoples of this hemisphere and some from the invading conquistadors and other early European settlers.

My favorite Native American name for these cats is Ghost Walker. These cats can appear and disappear in a flash; they are ambush hunters. It is the perfect name, as Sasha taught me more than once.

Sasha's pen was sixty feet long and ten feet wide with a

guillotine door (a guillotine door is a small door in the middle wall opened by a pulley system). The door allowed a human to close her off in one end while the other end was cleaned. I never used it. Since Sasha had taught me right away that we could respect each other's turf, I made my own call on that—and kept my guard up.

On more than one occasion I went out to feed and play with Sasha and couldn't find her. At first I would check to make sure the doors were locked and the fence intact. Then, with no big cat in sight, I'd begin at one end of the enclosure and look at every square inch. No Sasha.

I wouldn't enter her enclosure without knowing where she was, and there were big trees and big logs so she had places to hide. From the outside, I'd check around and under every log. Still no Sasha. I'd wonder if somehow she had gotten out, and then I'd think no, she can't have escaped, she has to be in here. I'd start from the other end and go even slower, looking everything over inch by inch.

Finally, I'd squat down to look under logs. Still no big beautiful maddening cat. I could only marvel at her vanishing acts. She melted into the setting so well that even when I looked straight at her I couldn't see her, rock still. One time when I was going through the same routine more than a few times and not finding her, I saw something flicker next to a stump. It was a green eye. I had her. She had a huge grin. Ghost Walker, Ghost Trickster.

❦ ❦ ❦

MY FOLKS LIKED THAT I volunteered at Sarvey. They lived in San Diego, so we didn't visit often, but Lynda and I kept them up to date on what I was doing. I'd told my parents Sarvey stories and sent them photos of my animal friends. It was a shock in May 1996 when my dad called to tell me that he had been diagnosed with multiple cancers. I couldn't imagine my world without him. He was a retired marine, a veteran of two wars who still wore his hair in a military crew cut. I wanted him to be able to visit Sarvey and see our work while he still could. My dad, who had been the first person to teach me to love the natural world, had seen my pictures of Sasha and wanted to meet her.

My folks arranged a trip up to Sarvey in the spring of 1997. I took them on the grand tour. We started out with the two-leggeds. They met Crazy Bob, who was always on his best behavior meeting family of people he liked. Kaye went on the tour with us. I took my parents to the enclosures that held the bobcats Bear and Grandpa. We visited a few raccoons, the eagles, owls, and Cetan, a red-tailed hawk who was the very first raptor I had ever handled with a glove.

My dad liked the tour, but he really wanted to meet Sasha. We all went out back to Sasha's enclosure.

"I'll stay back, so Sasha won't get upset," Kaye said.

My mom gave me a look. "It's just Kaye she hates," I said. "They're both dominant females." If Kaye walked by Sasha's enclosure, we could see just how scary a cougar could be. Sasha would charge at the chain-link fence, stand up on her hind legs to her full height, and slam the fence with her front paws. Her ears would be flattened, eyes glowing green, teeth bared—and she'd be hissing the whole time.

When we came up to the enclosure, Sasha was lying down facing us on a tree stump in the back of her pen. I said to my dad, "Why don't you walk up to the cage and talk to her?" But as soon as he got close to her, she turned her back on him.

My dad turned around with a baffled look on his face. I looked over at Kaye, who was standing at a distance with my mom. We both laughed. "He's an alpha male," Kaye said to my mom. "Sasha's a dominant female. That's why she turned her back. It's a sign of submission among cats."

My dad was excited to be close to her. He knew how I loved Sasha. I took him inside her enclosure. Sasha came directly up to me and turned to my dad. Her trust was instantaneous. Her motor was rumbling. I could tell she wanted to play. He scratched her behind the ears and around her neck. She let him rub her chin. He had the smile of a delighted kid.

THROUGH 1996 AND 1997, Sasha and I became almost litter-mates. Crazy Bob took even greater risks and committed even greater kindnesses. Kaye and I developed the kind of fiery relationship that two strong-willed people often have. While our shared dedication to the wild ones was never questioned, at times we disagreed about how to best manage Sarvey from a financial and upkeep point of view. And when we disagreed, the sparks would fly. We fought and made up, fought and made up again. And, each time, we always came back to agreeing on what would be good for Sarvey.

And then, on August 12, 1998, I met a new arrival who would sweep me further down the current of my life.

CHAPTER FOUR

CRAZY BOB HAD BROUGHT the young eagle in. She had been found at the base of her nest-tree, a tall Douglas fir. Bob said that she was one of a pair born to parents who had nested in the tree for eight years.

While I held her Kaye checked her for broken bones. We administered subcutaneous Lactated Ringer's, a solution that contains electrolytes and more. Then we weighed her. She was emaciated, weighing in under eight pounds, and covered in parasites. She was dusted for lice.

I loaded her up and took her to the vet. My old Escort had no room in back for the big dog carrier we had put her in. I pulled the passenger seat out and set the carrier in with the top off so the eagle and I could see each other.

I think it was on that thirty-minute drive that our connection began. We talked all the way down, me with my voice, the eagle with her eyes. I spoke softly. "It's going to be OK. We'll get you fixed up." She stared at me with her huge dark eyes. She was gorgeous, with the rich hues of her brown feathers and the way her beak shone. It was ebony—the richest and deepest I'd ever seen.

At the vet's, I gently lifted her out of the carrier and placed her on the x-ray table. She didn't struggle—she just let me take care of her. I knew, from working with other injured eagles, that she had been through so much that she didn't have any fight left.

The vet sedated her, then took x-rays of each wing along with full body shots. Once the x-rays came back and we knew the extent of the damage to her wings, the vet put a stabilizing pin in each wing. The pins were threaded stainless steel rods. He inserted each pin through the joint into the pneumatic humerus bone of the wing. A pneumatic bone is a hollow bone for air conduction; it's called an air exchanger. All birds have hollow bones. It takes two eagle breaths for the air to completely flow through. The air conduction is critical for flight.

He left part of the pin sticking out for ease in removal. Her feathers were still growing so we didn't want to impede that growth.

The vet wrapped the eaglet's pinned wings. We woke her by giving her oxygen and waited to make sure she was stabilized. Then I put her back in her carrier in a nest of shredded newspaper and drove slowly back to Sarvey. I asked her how she was feeling, and I told her she would have something to eat and a warm, safe place to sleep.

I knew from working at Sarvey that when a bird gets so thin that you can feel her keel (a bone that is attached at a right angle to the breastbone) almost as a sharp blade, the bird is in serious trouble. During intense starvation, the muscles start consuming themselves. It's very hard to reverse. In fact, once the muscle attrition begins, the bird is usually close to death. This eagle's keel was sharper than I'd ever felt before.

Once we arrived at Sarvey, the staff and I left her in her carrier-nest in the wooden dowel cage. She was awake, but quiet and still. Later that evening, we gave her more Ringer's. I silently asked her to give me a sign that she wanted to live.

There was nothing I could do but ask her to keep fighting. I would help her.

I couldn't go back to Sarvey the next day because of work, but I called and checked in. She was in essentially the same condition. I hadn't really expected anything different, but I knew that if she didn't get on her feet soon, it would indicate she was in pretty serious trouble. It's hard to tell the prognosis with a young eagle. There are so many variables—but will to live is a strong factor in recovery.

She started on raptor tube mix in the next day or two, because when raptors arrive emaciated, we can't feed them solid food right away; it could send them into shock and kill them. It is the same as with starving people.

I called every few days. One of the many volunteers would fill me in on what was happening. Since volunteers work different shifts, I'd often speak directly to Kaye. I'd have to wait till Thursday to make my regular visit to Sarvey, but I couldn't wait till then to hear how she was doing.

The news was always the same. She was alive, had graduated from the tube feedings to chopped-up rats, quail, and beef heart, was eating just fine, and was not standing up. We did have to make sure she didn't get any fur or feathers with her rats and quail. Normally you want an eagle to have that, but in her weakened state we didn't want her to expend the energy to digest the meat from the bones and feathers, and cast a pellet. She needed every ounce of strength she had.

At last Thursday came. I felt a little worried but not much. It was too soon to really get concerned over her fate. I knew after three years of Sarvey work that what would be, would be. When I arrived at the center, I found the eagle lying on her belly in her carrier. She looked up at me with her bright eyes. A volunteer told me she was eating more and more each day. It had been a week, and she was gaining weight, but making no progress in standing. I knew from working with other raptors that if she didn't stand, it was a sign she wasn't going to make it.

I reached into the cage and picked her up. I hoped she would be able to stand if I set her on her feet. I braced her, but no luck. I tried once more and then let her lie back down. I leaned in close and said to her softly, "You gotta get up, little sister." She looked at me with her obsidian-dark eyes. I wondered if she knew the right pace of her own healing.

For two weeks, it went on like this. I called from work and got the same news—she's eating, gaining a little weight, and lying on her belly. I visited every week, sometimes once, sometimes twice. I sat with her, picked her up, and talked to her more and more. Each time I came I was more concerned for her future. I knew that if she didn't stand pretty soon, we would have to think about euthanizing her. A life without mobility for a wild one is torture for them and ultimately, because of muscle wasting and general attrition, a death sentence.

I was discouraged each visit and in the long hours between, but I dared not show it to the eagle. She was a kid in the fight for her life, and I didn't want to add to her massive stress. Lynda was my only confidante beyond my friends at Sarvey, and all anyone could do for me was listen.

When I visited, I would look into the eagle's eyes. They truly were windows; you could see the confusion and fear—but you could also see the infant wonder. She was clearly soaking up everything she saw around her. She was learning day by day, watching us care for her.

I knew her physical signs were one way to judge her progress. And those signs weren't good. Though she was eating, she was still reeling from the physical and displacement trauma, and she just didn't have the strength to be able to stand.

The will to live is harder to measure. Working with wildlife you can develop a feel for that. I was searching for some hint of her spirit to survive.

Three weeks down the road, she was still on her belly. I tried to remember what Kaye had told me about life force. She studied a lot of spiritual ideas and she'd been talking about "mana," which is an Eastern understanding of the spiritual center of gods and holy ones. I wondered if that's what I had felt in this eagle right from our first gaze to gaze. *Look* isn't a strong enough word for that eyeball connection. I felt like I was right down inside that other pair of eyes.

On the third week visiting with her, as I was sitting with her and talking with her, I was trying not to think, *Are you dying?*

I'd seen many deaths at Sarvey since the deer on that first day. I'd been present at the euthanizing of more than a few raptors, mammals, and birds. I thought of what Kaye had told me about when her mom was dying, and how Kaye gave her permission to die. "I told her it was OK to go. Then I fell asleep for about half an hour. When I woke up, my mother had died. She needed to be alone to go."

I looked into the eagle's soft bright eyes and I thought, *No way, little sister. I'm not giving you permission to die. You do what you gotta do to live—rest, heal, whatever it takes.*

Day after day, week after week, a full month it went on like that. She had been through so much with the fall, then lying nearly dying on the ground, then the nightmare of being treated, trapped so far from her nest and her parents. Sometimes I'd see the spark brighten, other times it seemed like her mana was almost dry.

Four weeks, five, and by week six, the eagle was still flat on her belly. The staff began to think about euthanasia. I didn't really want to hear that, even though I know that rehab should never cross over to torture. This eagle wasn't being put through any painful treatments. But for the Queen of the Sky to be trapped in her body had to be brutal.

By now I had grown deeply attached to the eagle. Instead of discussing what to do, I'd go back to where my eagle sister lay and I'd talk to her. I hoped sending positive thoughts would help her, since it was all I could do. We were just getting to know each other.

She was eating fine and gaining weight. Her eyes were bright and clear, but she was not showing any effort to get up. In the seventh week it seemed very grim.

I can still remember the looks on Bob's, Kaye's, and Judy's faces during our next conversation. I knew mine looked the same. Finally Kaye said, "We have to make a decision. Soon. How long do we let this continue? It's no way for that eagle to live—or die—lying on her belly day after day."

Together we decided that if the eagle was not up on her feet in a week, we would have to euthanize her. I left my friends and coworkers with a heavy heart. But I wouldn't let the eagle know that.

She looked up when I approached the dowel cage. "Hey,

little sister," I said quietly, "now is the time. If you want to live, you've got to stand up." She looked up into my eyes. I couldn't tell what was going on behind them. I reached in and gently lifted her. Not much happened. I tried once more. Again nothing, so I just sat with her for a long time.

I drove home feeling empty. And I knew the only sight that would fill the hole would be to see her standing.

I called every day for the next six days. I felt more and more jangly every time I heard the words, "She's still not standing." Each day I began to dread the phone call of the next day. No matter what I was doing—driving to work, trying to concentrate on the computer screen in front of me, driving home, trying to watch TV, all I could think about was the possibility of that bird losing its life.

Then it was Thursday. I did not want to go to Sarvey. I knew I couldn't face seeing her on her belly and knowing that it was over. It would be exactly like looking at her dead body. I made myself get up, grab some coffee, and get in the truck before I could change my mind. If the eagle had had the courage to stick around as long as she had, I owed it to her to be there when she died.

The drive over was a nightmare. I just kept focusing on the road and each mile at a time. Normally I love this drive, and I see many wild ones. But on this day I couldn't see anything.

The turnoff to Sarvey came too soon. I pulled up the dirt

drive, parked, and just sat. I fought it out with myself. I was dying to turn around and drive home, but I said to myself, "Don't do this. Get in there. She's your sister."

Finally I sucked it up and walked to the door of the clinic. It was both the shortest and the longest walk of my life. I went in. Judy was at the computer. A couple of volunteers were chopping veggies and beef heart for the wild ones. Everybody looked up and gave me a shit-eating grin. There's no other way to describe it. Nobody said a word.

What the hell is going on? I thought. I didn't bother to check it out. I was in too much of a hurry to be with the eagle.

I had to walk through the clinic kitchen to the door to the cage room. As soon as I stepped into the doorway I could see the dowel cage and the most beautiful sight imaginable—a young eagle standing on her feet in the corner closest to where I would come in. She was craning her neck so she could see as far as possible. I knew she was waiting for me.

I felt my own grin spread across my face and raced over to her. "Little sister," I said, "it's about time. Good to see you up." I was about as happy as I've ever been. My eyes might have even been a little watery.

I opened the cage door, reached in, and scratched her soft belly. "Way to go, kid," I said.

Judy came in. "She stood up about an hour ago. We knew you were driving over and we had no way to reach you. Besides, you know we love this kind of surprise."

After another week, Kaye asked me to take the eagle to the vet to have the pins in her wings removed. I talked to her on the way down, telling her she was past the hard times and that this wouldn't take long. She was perfectly calm the whole trip. The vet saw us right away.

I put a sheet over her head so she wouldn't startle and held her while the vet did the work. He unscrewed the pins and pulled them out. Her right wing was perfect, but her left was not. She couldn't fully extend it. The wrist was frozen, and deep down I knew it was beyond hope. I put her back in her carrier and drove us home.

Everyone at Sarvey hoped a little time and physical therapy was all the eagle needed in order to heal enough to try to fly. But she made no significant progress after the pins were removed. Her left wing was too badly damaged. Her most serious break was near her wrist. She would never fly, never soar the skies with her people. Her life was saved, but for what? Was she doomed to live her entire life in a cage?

A cage can be a meager existence for her kind. Bald eagles normally want nothing to do with humans and will go to great lengths to get away from them. Even if they are used to people, they are mostly responsive only at food time. They are often aloof. In wildlife rehabilitation it's important not to get wild ones who will be released habituated to humans. But this eagle could never be released. And she was different from most eagles—she liked people. Her bright eyes followed most anything that moved. She wanted to see what we were doing.

About this time Kaye suggested that I try to glove train our new resident. Glove training is a series of small mutual familiarizations that lead to a raptor willingly stepping onto the thick glove the handler wears. The bird can then be carried on the handler's arm. Our new eagle had the right temperament. She might even have the equanimity it took to do educational programs.

I knew it would work. My eagle sister was more than trainable—she was quick and curious. We already had the bond that would be necessary. We had the mutual trust. We had the mutual responsiveness. Although few eagles are willing to be handled, much less remain calm in front of large crowds, I knew our girl had the disposition and the curiosity to be on display. She deserved a chance to show her stuff.

Our training began.

CHAPTER FIVE

As soon as I began working with the eagle, I felt a knowing, not in my mind, but inside my being. There was an ease, a calmness with this bird—and it was just between me and her. I'd truly never felt this with another animal, no matter how close I was with him or her. And I sure hadn't felt it with myself.

I was amazed every time I looked down at her talons and remembered how powerful and gentle they can be. Up close, the yellow scales on the top of her feet look reptilian. They're small and dense up by her feathers, then wider across her toes.

The talons could rest on my arm gently—and close lightning fast with incredible strength. I knew that the eagle could take me down to my knees with those talons. If she wanted to. But she didn't want to.

I began glove training in October 1998. My initial attempts were more like a comedy routine than formal training.

At first, our training sessions had a pattern. I would put on a heavy leather glove that went to my elbow. I'd go to the dowel cage and open the door. The eagle would scoot to the back of the cage. I'd sit down on a stool in front of her cage. We'd look at each other for a few minutes. I'd slowly reach into the enclosure and set the big leather glove on the floor.

She would eye me with perfect skepticism. "What's that thing on the floor of my cage? What do you think you're going to do with it?"

I'd gaze right back, suppressing a grin. "This is a glove," I'd say quietly and put the glove on. "All you have to do is step on it."

"Excuse me? I don't think so!"

"C'mon. Just try it once."

The first time she stepped with her right foot on the glove, she had a look in her eyes that said, "OK, pal. Now what? This better be good." And then, as quick as I imagined that, she stepped back. "You gotta be kidding," her look said.

I put the glove near her right foot and touched her with it. She stomped on it like it might be prey, then with a thoroughly confused look, she stepped gingerly on the glove— then jumped back. I was hunched into the cage, reaching way back to where Her Highness was eyeing me coolly.

We did this for twenty to thirty minutes at a time, then

we would take a break and start all over again. She'd put her left foot on the glove, jump back, put her right foot on the glove, jump back, touch the glove again. Sometimes she would reach down and touch it with her beak. That was a beautiful moment—she was an inquisitive child, ever so gentle.

I had a hunch she was thinking, *OK, I'll step on your hand but only with one foot.* Then, *OK, I'll use both feet but only for a second.* Another week later, *Yeah, you can take me partway out of my cage, then I'll jump right back in.* And finally, after three weeks, *OK, I'll let you walk around with me on your arm.* And a few minutes later, *Hey, this is fun!*

As I got to know her, the need for a name became obvious. I wanted a Native American name for her, but none we came up with struck me as right. I wanted one to describe who she was. A few of us were hanging out at Sarvey a few weeks into training and somebody said, "What should we call that eagle?" Paula, a longtime volunteer, said, "How about Freedom?"

"Freedom" might seem a strange name for her. She is an eagle who will never fly. But as the days and weeks went on, everyone called her Freedom. It's never felt altogether *enough* to me, but at least it evokes her indomitable spirit.

A few weeks after she first let me walk with her, it was time for her final test—anklets and jesses. The anklets are leather straps loosely secured around her ankles, and the jesses are leather straps that slip through her anklets and attach to a leather leash. I put the anklets on her—a piece of cake. Then the jesses. It was as if she had been born with them on.

The ease and trust were remarkable. We were partners. I understood that we were teaching each other, and we were working together in a way that was most unusual.

Many raptor handlers use very different methods to train "their" birds. I never felt that Freedom was "my" bird. I was her human. And where other handlers might use cruel methods of coercion, at Sarvey we never do.

If she didn't want to take part in something, I sensed it right away—her body language was very clear—and I didn't push it. Our relationship was always new, always changing. Sometimes she could be a stubborn child. And I learned over time to sense how she felt. For example, early on I got to know when she didn't want to get jessed up. And when she wanted to try and fly and she couldn't, I shared her frustration.

Soon most of the physical steps necessary to having her on my arm and carrying her became second nature. Going from her dowel cage to the outdoors required moving through some tight spaces, past a few volunteers, around a corner, and through a couple of doorways. Raptors have a behavior called *bating*, when they try to fly off the handler's arm. Bating in the doorway is particularly problematic—she could seriously injure herself by hitting her wings on a steel cage or the door frame.

I learned early on to be thinking three and four steps ahead of us as we moved. I would jess her up while she was in the dowel cage. She would step on my arm and I'd bring her out. She'd flap her huge wings, stretching them out, many times batting me about the head. I would wait for her to calm down before heading outside. As we approached each doorway I would close her wings up by wrapping my left arm all the way around her in a loose hug. She wouldn't struggle. Eagles do not normally allow this, but Freedom always allowed me to restrain her in this gentle but firm way.

I'd slowly walk her away from the dowel cage. She would be calm, but the closer we came to the door, the more I could feel her excitement building. Her muscles would twitch as adrenaline surged into them. She'd crane her neck to look ahead, and I'd feel her heart pound. I'd feel my own excitement for her too. By the time we were outside the last door, she was ready to explode. I'd let her out of my embrace and she'd move into a full bate, a "get me outta here" launch.

Her eyes were wide. Anyone could see the primal instinct in full force at that moment. Every sense she had was in overdrive. She was ecstatic to be out under the sky. And I was almost as happy to be able to give her that small liberation.

Besides the glove training, we had to get her ready to move outside, but first she needed to be waterproofed—gradually exposed to water so that her natural ability to repel it would develop—and she had to put on a little more weight.

She was a finicky eater from day one, very un-eagle-like. Eagles in the wild eat very fast; they can't afford another eagle stealing their food. Most retain this habit even in captivity. Not Freedom. She looks at her food and often enough turns away, returning to her sentry post to survey her domain. I don't remember a time when she didn't have this behavior, so I don't know if while she was sick she learned that no one would take her food away at Sarvey, or if she never had the instinct to protect it.

She has never been a voracious eater. She exhibits another behavior many captive birds don't—she gives herself a bye day. For a bird of prey, a bye day is a day of not eating. Raptors living in the wild do not eat every day, because hunting may be unsuccessful. Many handlers give their birds bye days to simulate behavior in the wild. At Sarvey we use bye days in the summer and fall seasons to keep weight consistent, but we've never needed to do that with Freedom because she gives them to herself. She also rarely does it during the molting season, when we never give bye days to a raptor, since making quills and growing new feathers takes a lot of energy.

But when it comes to baths, Freedom's response is pure instinct. One of eight species known as sea eagles, bald eagles are fishermen; they have been known to catch salmon that are much too big for them to fly with and they will swim them to shore. They are fastidious birds and love the water.

The first time I gave Freedom a tub of water was party time, no hesitation. She dipped one wing in, then the other, submerged her head, all the while fluffing and shaking her body, water flying everywhere. In front of her dowel cage, it looked as if somebody had turned a hose on and let it run. Everything was drenched.

Freedom had missed the chance to live outside in the winter. By the time she was healthy enough to start living outside, the near summer days of late October and the cool autumn nights had given way to the dark cold days of November and December.

Eagles acclimate gradually to the cold by living outside in the fall.

We wanted Freedom to live as much like an eagle as possible and be outside all the time. So when the spring came, I introduced her to her very own flight (outside enclosure). I got her up on the glove, and we went out back. I took her into the cedar flight hallway and then into her new home. She bated as soon as we got into the flight—just like an excited youngster.

I set her on the main perch and took off her jesses. She scoped everything out. I had taken some beef heart out with me so she would know that place was hers—her food, her perch. I set it on a stump, but Freedom had more important things to do. She jumped down and went exploring, checking out one corner and another, and then she saw it—the pool!

All her very own. She did her excited loping eagle hop over to the pool and stared into it. All was right in her world; the sun was shining, and she had room to move and there would be a bath.

Before Freedom was ready to be in education programs, I needed to get her used to travel. That meant her willingly getting into a large carrier with a perch in it. Some birds don't like that. A few weeks before our first program, I decided to get her used to the carrier and the truck. First I got her transport carrier ready by putting newspaper on the floor and her perch inside. Then I got her all jessed up, and out the door of her flight and down the hall we went. We got out front and I really didn't know what to expect as I brought her near the carrier.

As soon as she got within three feet of the carrier she jumped in, stood on the perch, and was ready to go—showtime. I didn't know how to tell her this was just a rehearsal. She was ahead of me. *So, OK, let's go*, I thought. I closed her up in the carrier, shut the truck canopy and tailgate, and hopped in. Off we went down the hill and down the road. We took a fifteen-minute drive and came back to the center. She was standing tall on her perch as if she had done this a thousand times before. I was so proud of her I was busting at the seams.

Her first program was an Eagle Scout ceremony. I got her ready to go, and, atypically, I brought a pair of eagle gloves.

Usually I just took one, for the hand I used. All this time I had been using the left hand as all falconers do. But when we got to the church where the event was held, Freedom started getting antsy. I couldn't figure out why she would be nervous.

I got her out on the glove as Kaye and I waited to go on. Our time came, and for the very first time, Freedom was out in public in her big debut.

It didn't really go the way I had planned. Freedom took one look at the crowd and spun around so she was looking right at them with her back to me, her injured wing hidden from the crowd. "Uh—excuse me," I told her, "you need to turn back around and face me."

"No, I don't," she replied, and started fidgeting around on my arm.

She would not stay still. I could see she was getting ready to bate. *Great*, I thought, *I have an unruly eagle here in front of more than a hundred people.* No matter what I did, she wouldn't calm down. I gave up trying to turn her around, and we finished the show like that.

I put her back in the carrier after the show. One of the Eagle Scouts asked for pictures with Freedom and his family. I was hesitant because of the way she was acting, but just before I took her out I had an idea. I would try the right-hand glove. Maybe she didn't like the left when there were strangers around—maybe she wanted her damaged left wing close to me. In the wild, eagles can't show weakness; even their own kind will steal a weak eagle's food or attack them.

I put on the right-hand glove and reached in the carrier. She stepped on, and I brought her out.

She was completely calm on the right side. From then on I used my right arm. Freedom weighed about ten pounds, which might not seem like much but is a lot to keep on anybody's arm steadily. It's like holding a ten-pound bag of sugar in a clenched fist with the arm at a rigid ninety-degree angle to the person's side. In short order, the muscles in my right arm felt different from those in my left arm. "Eagle arm" I liked to call it—that was far cooler than tennis elbow.

EARLY IN 1999, SOME people from the North Cascades Institute—whom I knew from our days in the Eagle Watchers program—called and asked Kaye and me to bring Freedom up to the ground-breaking ceremony for their new environmental learning center. The center would be built near the Diablo Dam on the Skagit River. The ceremony was in June, and it would be the largest gathering Freedom had ever been to.

The day of the ceremony dawned bright and beautiful. I went to Freedom's flight, jessed her up, and transferred her into her carrier. Kaye jumped into my truck and we were on the road.

I would have bet there were more shades of green in that river valley that morning than anywhere on earth. The scent of pine and cedar filled the air. Kaye and I could hear the Skagit rolling and tumbling to our right.

We arrived at the east side of the dam and were escorted directly across it. All others had to take the long way around. We had a big sign we had to put on the dash, VIP. I said to Kaye that it should have read VIE.

"We're bringing royalty," Kaye said with a grin. I parked at the site of the ground breaking. Kaye climbed out. I wasn't ready to take Freedom from the truck; I needed to get her perch set up on her carrier first.

Some musician friends of mine from way back, David and Douglas Farage, owned an ornamental iron fabrication business and they had made Freedom a custom-designed eagle perch that fit on top of her carrier. I set up in the shade of a tall old cedar right by the dammed Skagit. There would be a few speeches by local officials, then the audience would explore some of the temporary displays that told the history, intention, and building plan for the center. Freedom was part of the education event. I held her on my arm as we waited for the program to start. Visitors greeted her. Everything was going as planned.

There were no clouds in a brilliant sky. A breeze washed up from the deep blue water. Shafts of sunlight pierced the deep green canopy. Suddenly, Freedom started getting agitated. She flapped her wings violently and bated. I got her back up on the glove. She yanked on the jesses, shook her head from side to side, and cussed me out. Her eyes were wide, and her crown was standing up.

No matter what I did, she repeated the performance again and again. I tried to put her on the perch. Freedom would sit there for a minute or two, then bate again. I couldn't figure out what was happening. Was there a human she didn't trust? Was it the setting? I had no idea.

I put her back on the glove. I put her on the perch with the same results. I tried everything I knew to calm her down. And then it dawned on me.

I thought back to my years with the Eagle Watchers program on the Skagit. A year or so after my introduction to the program, I got lucky and witnessed a display of speed and power I had never seen before.

It was on a day as brilliantly sunny as this one—but I was alone. I spotted half a dozen eagles flying east on the northern ridgeline above the river valley. Suddenly, one bird banked away from the rest, making a hard right turn—almost turning over on his back—then dove and barreled down the long steep slope just above the tree line. I was watching him up close through binoculars and I could see him bathed in sunshine, feathers ruffling in the wind as he dove. At the last minute, he made a violent U-turn and hurtled back up the slope, then doubled back heading toward the river behind me.

I made a guess, based on what I knew about eagles and the speeds they can reach, that this eagle was probably doing at least sixty miles per hour.

Right after that I saw twenty eagles or more gathering overhead. When eagles ride the thermals together in that way, it's called *kettling*. They can be seen from miles away. They catch the thermal's warm up-current, spread their wings, and soar higher and higher. In many native beliefs, when the birds disappear from sight in that way, the eagles are delivering prayers to the Great Spirit.

I knew that Freedom felt in her bones the same thing those eagles had felt. She recognized this place as where she should be, where her kind live. The wild was calling this wild one. The wild was calling Freedom home. I could see she was frustrated and agitated, not being able to fly. I knew what I had to do.

I turned to Kaye. "I am not subjecting her to this. We're done." I put Freedom back into her carrier. I explained briefly to the officials what was happening. They understood. Kaye climbed back into the truck and we left.

Kaye looked at me as we drove away. "Good call," she said.

I learned that day to always judge situations in advance so Freedom would never be in that position again.

❦ ❦ ❦

A COUPLE OF MONTHS after the ground-breaking ceremony, Freedom, Kaye, and I started going to local Native American powwows. Freedom had basically three jobs as an education bird at powwows and other programs. She would serve as a teacher about eagles, their place in the natural world, and the importance of protecting their habitat. She taught in a way that no book or slide show could. Anyone could talk to kids, show them pictures, and still they'd be bored, but when I brought a living eagle into their room, we had their attention.

Freedom was the public relations eagle for Sarvey—the face of the center to many. She brought recognition and crowds. New volunteers for Sarvey sometimes signed up after meeting her.

And Freedom was an important part of what is called Grand Entry at the beginning of a powwow. A powwow is a dance celebration and competition held by one First Nation, but including dancers from many nations. The eagle staff carrier would lead the procession into the powwow circle before the dancing began. Kaye and her red-tailed hawk, Mellow Yellow, would step out behind him, then Freedom and I would precede the procession of dancers, flag carriers, and military veterans into the powwow circle.

As Freedom and I took part in the powwows, I began to learn more. The First Nation people that I encountered in their formal roles at the powwow, and later as friends, began to teach me about their ways. They didn't lecture. They treated me as an equal.

Many First Nations revere the eagle. One of the Coast Salish people's stories tells of a great flood. Everything except the highest mountains was under water. Only one of their people lived. He had a dream that guided him to escape to the top of Mount Chuckigh. Then a huge eagle came to him carrying a salmon to eat and told him the flood was gone; he could go back down.

When the man descended, he found that all of his Salish people were dead except for one woman. He married her. To show their gratitude to the eagle who had brought him food and the message of hope, the couple took the eagle as their chief totem. The Coast Salish people to this day honor the eagle in song, ceremony, and prayer.

Even farther north, the Athapaskan people who live in forests below the Arctic Circle saw the eagle as a lifesaver for their people. One of their leaders had given a salmon to an eagle. Years later, when there was little food, eagles brought the Athapaskans not just salmon, but sea lions and whales for their hunger.

At these powwows, prayers included everything— humans, the ones that crawl, the winged ones, our rooted friends, the four-legged and two-legged ones, and the ones that swim. I learned more about the importance of honoring human connection with the earth and all our relations.

I learned to trust something greater than myself. I learned to trust something that I had no name for, but my teachers called the Great Spirit or Grandfather. A few months into my work with Freedom, I learned more about trust.

MY DAD HAD BEEN battling cancer for a couple of years. I knew his time was short, but it was still a jolt when my sister, Jill, called me at work and said, "You need to come home. It's time."

Lynda booked me a flight and saw me off at the airport the next morning. We were driving just down the street from our house, when something caught my eye. "Look at that," I said. "I've never seen a red-tailed hawk there."

"Me either," she said. I slowed a little so we could check out the bird. A big red-tail was perched on a light pole right before the highway ramp. We were running on a tight schedule, so I pulled onto the highway without stopping.

I caught my flight and arrived in San Diego in the early afternoon. Jill came with my uncle Boyce and aunt Ann and drove me right to the Hospice Center. As we walked up to the building, I deliberately hung back from the others, lost in thought, taking my time.

They were about sixty feet or more ahead of me. Everything, the air, the sky, the people ahead of me seemed as bright as if they were made of glass.

There was a sharp drop-off down a cliff to my right. Just then I saw a red-tailed hawk flying straight toward me a little above head level. He wasn't working hard, just cruising, the sun shining through his red tail. Three crows flew in a tight circle above him, keeping pace with him. I couldn't believe my eyes. I stopped and watched.

All the birds came even with me.

And then suddenly the hawk was gone.

Time stopped. I looked all around but no hawk. There was nowhere for him to disappear to—and he had vanished.

"Hey, you guys," I yelled. "Did you see that?"

My sister, aunt, and uncle turned. "See what?"

I know that Native Americans have long believed that a bird can represent the soul and that the eagle and hawk can help with the passing from one world to the next.

We went into my dad's room. Jill's husband, Greg, was already there. My dad was barely alive. He was breathing that weird choppy Cheyne-Stokes rhythm—no breath for a long time, then a sharp gasp. It was just a matter of time. Greg said softly, "You want to be here by yourself?" I barely managed to nod my head yes.

I said my good-byes, but I knew my dad and I were already cool with each other. Nothing was left unsaid. That evening, my mom, Jill, and I stepped out onto the balcony outside the room. It was dark, a little after seven. Only the light from a few windows illuminated the bushes below. It was peaceful. We were quiet. Not much to say.

And then my mom almost yelled, "Did you see that?" Jill and I looked at her. "A red-tailed hawk just flew out of those bushes."

"No way, Mom," I said. "Hawks don't fly at night and they sure don't fly out of bushes."

Before she could say anything, my dad's sister Helen burst through the open door and said, "Quick! Get in here. It's time."

We went back into the room. My mom grabbed my dad's hand. My sister touched his arm. I stood next to my mom. Dad took his last breaths. And was gone. It was the first time I had seen a person die.

I was surprised at how peaceful the exact moment of my dad's death was. I remember thinking it was the most peaceful moment I'd ever experienced. There was nothing—in the best way—for me to cry about. And I will always wonder, maybe until my own death, what he felt at that moment.

I came back to Sarvey a few days after my dad's death and went straight to see Kaye. After our hellos, I got to the point. "Remember that you told me you were going to send a spirit bird with me on my trip?"

Kaye grinned.

"So what was it?" I asked.

"A red-tailed hawk," she said.

It was my turn to grin. Then she said, "Your dad died early Friday evening."

Now I don't know why I was surprised. "How'd you know?"

"The spirit hawk," she said, "was sitting in the tree that evening. I didn't see him come and I never saw him leave."

CHAPTER SIX

M Y DAD WOULD NEVER have wanted his family to stop living because he was gone. We had all promised him we would take care of one another. I had lost him, but I loved life. I really wished he could have met Freedom before he died. He would have loved that eagle. She gave me a sense of purpose in my life that hadn't been there before and that helped me stay focused on living.

My mom called a few months after my dad's death and told me that he had wanted to give money to Sarvey to build a big enclosure for Sasha. I remembered the look on his face when he'd been petting her under the chin. It would be a joy to honor his wishes.

Sasha's new home was five times the size of the old one and higher by at least five feet. It was built from chain link so that she was safe indoors and out. The old enclosure opened into it through a guillotine door.

It was only Sasha and me the first day she went into her new home. She was in the old enclosure. I went into the new one and called her over. She came and looked at me curiously. I raised the guillotine door and she sunk down low. "Come on, Sash," I said softly. She slunk in. As she moved past, I gently scratched along her back and let her tail run through my hand.

She did the slinky walk all the way around the compound. Her ears were up and her whiskers twitched. There were tears in my eyes watching that Firecat, Cat of the Gods, Ghost Walker slink around the perimeter and check it all out.

In less than ten minutes she was at home, walking tall. I knew my dad was with us at that moment. I knew he saw the joy in the faces of Sasha and me.

Sasha's new home was even better for hide-and-seek than the old one had been. When I went to take care of her, I'd go to the back door and start looking for her. I knew her strategies in her old outdoor cage, but now she had new ways to trick me. I'd go back and forth, front to back, side to side and there'd be no evidence of Sasha. Then, like as not, I'd see something twitch. It was the tip of a long tawny tail. I knew the twitch was her little way of telling me she had just kicked my ass again and the game was over.

Other days, I'd come up to the enclosure and see her

stretched out in the sun. I'd go in and she'd glance at me and not move. Then I'd go about my cleaning. I'd smear anchovy paste under some of the logs and on the sides of rocks. I'd crush cinnamon fine and rub it on other logs and rocks—this practice is called environmental enhancement. It gives wild ones something new to experience and breaks up the monotony of their daily life.

Between my work with Sasha, Freedom, and the other animals, I felt my life was maybe the best it had ever been.

And then one day in April 2000 I felt a lump on my neck. I figured no big deal. When I told Lynda about it, she said kindly, "That's your head." But it wasn't.

The lump didn't go away. Lynda insisted I get it checked out. I went to Sarvey on my usual Thursday and mentioned it to Judy. She said, "Let me see it." Her eyes got huge. She looked really worried, but didn't say much more, only that I should get it checked out immediately.

I saw my family doctor at Pacific Medical. He examined me and sent me to Greenlake Hospital for a CAT scan. I figured it would be no big deal because I was basically pretty healthy.

I was a little surprised when the doctor called and told me they wanted to do more tests. I couldn't believe there would really be something wrong with me. He sent me to Swedish Hospital. The doctor there said, "We're going to do a fine needle aspiration." *No big deal*, I thought again. *They'll grab a few cells. I'll be out, pain-free, in five minutes.*

Wrong. It was a big deal. Being jabbed in the neck with a very long needle repeatedly is a big deal.

They took the samples off for analysis. When the doctor returned, I asked him what he thought. "It looks like a lymph node to me," he said. I figured it was a little inflammation; maybe I picked up a germ somewhere.

I was sent back to Pacific Medical to meet with the surgeon-in-charge-of-not-telling-you-much. What she did tell me was that the results of the fine needle aspiration weren't conclusive, and it would be necessary for me to have surgery to remove the lymph node to be biopsied. This information was a very big deal. I didn't like hospitals, and I was sure I wouldn't like surgery.

I drove to Sarvey right after I heard the news. I jessed Freedom up and walked out to our favorite spot on the long green slope looking down on the ponds. When we stood on our hill and gazed over the river valley, the world was just us and where we stood, but even there, I felt a serious foreboding. It was as though some tiny whining machine in the back of my mind wouldn't shut up.

I did what I could to keep that whining muffled. I did my best to ignore it. So it was almost a relief the day I finally went in for the surgery. Lynda drove me down and waited for me to get checked in.

I woke up from the procedure. The overhead lights were dimmed for those of us coming out of surgery, and the air tasted of antiseptic. The machines made weird spaceship sounds. My nose itched, so I took off my oxygen mask. I could hear the nurse telling me to leave it on, but my nose itched too much. We had a back-and-forth thing—I'd take it off and she would put it back on my face, then we'd start over again. The opiates they gave had made me itch. It was a funny little junkie dance that nurse and I did.

When I could go home a few hours later, they wheeled me out to the car in a wheelchair. That really irritated me. All I could think was, *I don't ride in wheelchairs, only sick people do.* It felt like a complete loss of dignity. My mind was yelling at me and the staff, "I can walk! I don't need or want this thing, this wheelchair!" but hospital regulations insisted I needed one. Plus I was pretty loopy from the drugs, too loopy to actively protest, though I really believed I could have walked.

I squirmed into the car and that hurt big-time. I came fully awake. The painkillers were wearing off, and any movement seemed to take my neck with it, even if I was only moving my foot an inch or two. Groggy, I wondered if everything in my body was connected directly to that one side of my neck. I knew for sure I was learning more about pain than I had ever wanted to know.

Lynda and I came home. I hobbled in the house and took the Vicodin. The pill made me feel really sick. I decided to wait out the sickness and not take another one.

I made it through the rest of the afternoon sacked out on

the couch and went to bed early, tired from an exhilarating day of scalpels and mean drugs.

I woke after a few hours and tried to roll over. There was instant eye-crossing pain. My mind was spinning. I was on a whole new level of hurt and was wide awake in all the wrong ways. I cussed, gritted my teeth, and tried again to roll over. My eyes watered, but I kept trying.

It hurt too much. I had to have Lynda push me over. What was this? And what was it going to be? I couldn't even roll over by myself. All I could think was, *Damn, this sucks.*

Morning came. I thought, *If I can just get up and slam some coffee, I'll win this battle.* I did sit up, but it took awhile, and staying up was another story. I fell back on the bed and had to try again.

No matter how carefully I moved, I felt like someone had jammed a long blade into my neck and was twisting it slowly. I clenched my teeth, so as not to bite off my tongue when the fun started, then rolled and slid to the floor. I wound up kneeling by the bed. I was determined to stand up by myself. I mustered all my gumption, corkscrewed up, and fell back into bed with that same invisible person sticking the knife in my neck. *Screw it*, I thought, *I'll just lie here.*

The pain was bad, but what was worse was not being able to do the simplest of tasks. I knew that Lynda was there to help me, but I did not want her to have to take care of me like this. The loss of independence hurt worse than physical pain. Finally, I managed to get into the kitchen.

I sat at the table drinking coffee and shaking the cobwebs out, wishing I could get past feeling like hell or concentrate on anything else, but I was forced to sit with it. I think I knew even then that I needed to pay attention to all that was happening to me. I had never been in this position before, having to focus on my health and not being able to plow through some problem. I didn't like it at all.

I finished the last of the coffee and just sat. In that moment, I began to understand there really might be a long hard battle ahead of me. And the only feelings I could really feel were anger and frustration. I kept thinking, *I don't have time for this! I really don't have time for this!!*

A COUPLE OF DAYS LATER Lynda drove me back so they could check the incision, clean it up a little, and send me on my way. The doctor changed my bandage, and I got a brand-new one all the way around my neck, a great big white one that stood out like a beacon in the night.

Great, I thought. Another reminder of what I didn't want to deal with. By that time, I'd stopped saying "no big deal." All I could do was wait to find out the news about the biopsy results. That took about a week.

One day just after my birthday during the limbo time, I took Freedom out. It was a soft gray morning with shafts of sea blue piercing the clouds. We both needed and enjoyed our walks, but I think maybe that time I needed her more than she needed me. I hadn't opened up to many people that I might have cancer. I hadn't told people that I felt hellaciously uncertain and more than a little afraid. I have always been a private person.

On that hill, with my best friend on my arm, I could let myself feel exactly what I felt. I felt my fear of a hard future, of being trapped in a body and treatments that would feel alien. I felt, almost more than anything, that I resented having to go through this. I knew the truth—there in the cool fresh air and the soft light, I could feel something coming. And Freedom was there with me in the way she can be—as an ally, a soul singer, and as a reminder of a real future.

I RETURNED TO PACIFIC MEDICAL and the surgeon delivered her message: "It's malignant, Mr. Guidry. You have non-Hodgkin's lymphoma."

"Shit," I said. "I don't have time for cancer."

I thought back to that moment with Freedom. I'd known then. "Can you tell me, what are the odds?"

"Fifty-fifty," she said as though we were playing the ponies. That wasn't really the number I was looking for. I started to shut down. I could feel myself getting even madder. All I wanted to do was get the hell out of there. I left, barely saying good-bye.

I drove home and broke the news to Lynda. She had never let herself believe it could be cancer. Her shock and pain didn't come out till later. She told me that after I left she broke down and cried and that she didn't want to do it in front of me because I had enough to worry about.

I needed some space and I needed to take Freedom out. I also needed to tell Kaye and Judy, but not by phone, so I went to Sarvey and walked in and told Kaye. "Goddamn it," she said. She knew the reality of what I was facing—from her own battle with cancer.

The next day I went back to my family doctor. He suggested a specific hospital for the next steps. I said, "Nope, I want to go to Virginia Mason." I already knew that would be my choice because Kaye had gone there and told me the oncology center there was one of the best in the world. And Judy had told me that was the way to go.

Lynda knew somebody who had gone through his own Big Deal and raved about a doctor named Andrew Jacobs. Lynda found out Jacobs was available and made the appointment at Virginia Mason with Jacobs for the next day.

LYNDA AND I WENT into the exam room together. Dr. Jacobs and a nurse came in shortly after we arrived. I liked him right away. He was a Brit with a dry sense of humor and a quick smile. I knew that what we were going to go through wasn't exactly going to be *Monty Python's Flying Circus*, but I sensed in Dr. J. a real ally, someone who would tell me the truth and not sugarcoat it. It would not be just Lynda, me, and Freedom going through what came next. He'd be with us too.

Lynda had her carefully researched and thought-out list of questions. Dr. J. examined me after he checked out the results from the CAT scan, then he sat down with us. "You have stage 3 non-Hodgkin's lymphoma. It's one of the more treatable cancers."

There's the first decent news in a while, I thought. He then told us that they use the team medicine approach at Virginia Mason. Teams of doctors from VM, Fred Hutchinson, and the University of Washington review all the cases. "Here's what I think will be the most effective treatment for you," Dr. Jacobs said. "The way we treat this cancer is with a combination of cyclophosphamide, adriamycin, vincristine, and prednisone. It's called CHOP."

I asked him why he couldn't just cut the cancer out. He said, "You have cancer all through your lymphatic system, hundreds of nodes are affected, and you have a nine-centimeter tumor in your spleen."

He asked about other symptoms such as night sweats. I didn't have any of them. I said, "The only thing I noticed was the lump in my neck."

"You'll have an infusion session once every twenty-one

days. Compared to other treatments, this isn't so bad. You can do it through the outpatient center."

I asked a question I didn't want to ask, but knew I had to. "What happens if someone gets through all eight sessions and still has the cancer?"

"Eight is the maximum you can get because we've learned that if it isn't in remission at the end of eight, more infusions won't make a difference. If that should be the case, you would be a candidate for a stem cell transplant transfer."

"No way," I said. "I don't want to go there unless it's absolutely necessary." I'd read an article on stem cell transplants and they were some extremely harsh treatments.

We were all quiet for a minute. Then I said, "Am I going to lose my hair?"

Jacobs said, "Yes."

In my book, that was almost as bad as the cancer. And yet a small price to pay to save my life. I flashed on all the times as a teenager I had to get my hair cut when many of the other kids had long hair.

At Sarvey, dark humor is one way we deal with painful realities. "Cool," I said, "in that case I'll get my head tattooed. I've always wanted to do that."

Dr. J.'s nurse went wide-eyed. "Oh my god," she said. "You can't do that."

Dr. J. just grinned. I liked him even better. I asked him what he thought my chance of survival was.

"We don't give odds but 75 percent of the people who have this disease, and go through this treatment, recover," he said. "Really, Jeff, if you've got to get cancer, this is a good one."

So the surgeon who said fifty-fifty had been wrong—but 75 percent wasn't exactly the slam dunk I wanted. "What's next?" I asked.

Dr J. told me I needed a bone marrow aspiration to get a sample of the bone marrow to see if the cancer had spread. We made an appointment for that next week.

The day arrived, and once again Lynda and I drove to the hospital. I checked in and then they took me upstairs to a private room with a view of Seattle. *Not so bad so far*, I thought.

When I had told Judy they were going to do this, she told me to take every drug they offered. It was good advice and I took it. By the time the drugs had kicked in I was feeling fine. I had to lie on my stomach as they inserted a huge needle into my hip bone for the sample. I didn't feel a thing, just some pressure. After that was done, I nodded off and woke an hour or two later. Lynda was there waiting for me to wake up.

"How do you feel?" she asked.

I grinned. "Just fine."

A few days after that we went back to Dr. Jacobs's office. He told me the good news. The cancer had not spread to my bones. We caught it in time. "So when," he asked, "would you like to start treatments?" He pointed to a calendar.

"How about Friday?" I said. The doc looked pleased, if a little surprised. "Hey," I said, "is there any reason to wait?"

"Nope."

"Then let's get it going."

I didn't tell him I would have started the chemo sooner, but Thursday was my day at Sarvey. I didn't want the post-treatment side effects to interfere with my being able to work with Freedom.

Lynda and I talked a little on the way back about how we were going to approach this. She seemed calm and determined. I knew that when she got motivated, nothing got in her way.

THE NEXT DAY I went up to Sarvey. I jumped in my truck and I lit a cigarette. That might sound crazy, but I figured I was already in trouble, so what the hell, who cares. If I was in the 25 percent that didn't make it, I was going to enjoy what life was left to me. If I was in the 75—I'd quit. On the drive up, I kept thinking over and over again, *I do not want to do this cancer thing.*

I took Freedom out by the pond. It was almost enough comfort just to feel her weight on my arm. I felt my spirits lift. After we did our walk and I put Freedom back in her flight with a tasty rat and a quail, I cleaned out her pool and refilled it for the bath that was sure to come.

I walked over to Kaye's house and went in. We sat there talking for a while. She didn't give me the sad eye or say some make-nice thing like, "You're strong. You'll get through this." She rummaged around and found a book called *Getting Well Again* on healing visualizations. She had kept it after her own bout with cancer.

She opened to a page and said, "I want you to read this. Do the visualizations three times a day every day."

I glanced at the page and thought, *Yeah, sure*—and said, "I'll give it a shot." She shook her head. She knew I wasn't enthusiastic about the prospect. "We'll see," she said seriously. We said good-bye, and I headed home to Lynda.

I called my sister, Jill, and told her what I was going through. I asked her to keep the news to herself until I told Mom. I dreaded that call to our mom. I wanted to spare her the knowledge until the last possible moment.

Next, I called the owner of the company I work for. "I need to talk to you," I said. I told him I had cancer and was starting chemo and that I didn't know how many days I would miss from the treatments. Then he asked me, "What do you need to live on?"

I gave him a dollar figure and he said, "OK, we'll put you on salary and you come in when you can." I had good health insurance with the company, so I felt like I could relax about money matters.

I visited with Freedom on my regular Thursday, except

it wasn't a regular Thursday—it was the Thursday before I began the infusions.

That evening, I called my mom. She and I chatted for a minute, then I said, "I have some bad news and some good news. I've been diagnosed with non-Hodgkin's lymphoma." I heard her take a sharp breath. I said quickly, "I start chemo tomorrow and the survival rate is high." I asked her to stay away while I went through the treatments, and as much as it pained her she honored my request. I didn't want her to see me weakened from the chemo.

LYNDA AND I TOOK off Friday morning for my first infusion. "Let's go the scenic route," I said. She nodded. She knew what I meant. We both loved the route across the 520 floating bridge, past the arboretum, and toward the Japanese gardens. The road curled under an old stone bridge and wound its way down a tree-shrouded lane. As I drove, I could see golden pollen swirling in the spring breeze and smell the scent of new flowers and grasses wafting in the open window. There was life starting itself all over again as I went into the fight for my own life. I knew for that moment, I too would be renewed.

By the time we pulled into the hospital parking lot, I felt fairly calm. Looking at the trees and that old stonework had reminded me of where the real medicine lies—within nature and time. Lynda and I walked into the lobby and got in line. There were five intake desks with hefty wood dividers so we patients could check in with a little privacy. I scanned the room. Some folks were in wheelchairs, some had a look of terror, maybe knowing they might be on their final run here. I didn't let myself think about the possibility of my having a final run. Other patients joked around. Other patients were quiet, their concentration seemingly turned inward. Some wore wigs, some went pirate with brightly colored bandanas. All were in their own little battle, all facing it differently. I was still thinking "them," rather than "me."

The elevator doors opened. A guy stepped out. He wore the dreaded bandana on his head, and I knew he was bald underneath the bright cloth. I felt a little hollow. *Oh man, I'm gonna be that guy. I'm going to get chemoed. I'm going to lose my hair.*

I almost wanted to go straight to the check-in desk and say to the admission clerk, "Come on, man, let's get a move on." I made myself wait. We checked in and went upstairs to get blood drawn. The nurse asked me if I'd like to keep the stent in since I was scheduled for infusion later. "No way," I told her. "Once I'm done here, you pull that thing out of me." I hated those stents already. She asked if I wanted a port instead.

"Not until every vein collapses."

I had learned from Lynda's research that a port was a little plastic device sutured into a vein in your chest. The picture of it had creeped me out. Having one in my body would feel like an invasion.

We took the elevator to the twelfth floor where infusions were administered. A big gentle-faced guy sat at the check-in desk. He smiled the second he saw us, not some professional it's-going-to-be-OK smile, but a smile with real warmth. "I'm Jimmy. You can sign in here."

We chatted for a few minutes. He indicated one of the chairs in the waiting room and said, "You can wait there. You'll be called soon."

Soon couldn't be soon enough for me. Finally, I heard my name called, followed the nurse, and settled into a big green worn-in chair. I was ready to go.

The nurse asked me, "Which hand do you want it in?" *Neither*, I thought—or both of them so we can get to the end of this. "My left." She stuck a needle into a big vein on the back of my left hand and pulled the needle back out, leaving a plastic stent in the vein. I was then hooked up to a bag of chemo fluid. I made myself watch as she opened the release valve and adjusted everything to get the drip just right.

The poison—I was having a hard time thinking of it as "medicine"—began to move impossibly slowly into and down the tube. *Drip, drip, drip, drip.* I thought, *I am going to be here for the rest of my life.* I began to realize that I was going to be alone with myself a lot over the next eight months of CHOP. I also realized that I didn't really know what was going to happen to me over the course of the treatment. And it occurred to me that having to explain to other people what I was about to go through when I had no idea myself would be difficult at best and maybe even impossible.

I watched poison leach into my body, not really knowing how to fight for my life. How strange that sitting there doing the last thing in the world I ever wanted to do was part of the fight.

I checked the clock. Damn. Ten minutes had passed. And they had told me the session was going to last three hours. I hunkered down for the duration. I had two choices; it was read or people watch. I went with quietly watching. Everyone was in their own private struggle, doing the best they could. Not a lot of chatter went on there.

After about an hour, the first poison was in, and the second was about to begin. It was bright red, and there were two big plastic tubes of it. I reminded myself, *I'm fighting the enemy. I'm sitting tight.*

The nurse plugged the new line into the stent. I was fascinated watching her do this. She kept pushing the huge syringe in a little at a time, then pulling back to make sure the blood was flowing. I asked her why she did that. She told me that the red stuff was so toxic that she needed to make sure none escaped into the surrounding tissue. They even kept a syringe full of some type of steroid on the table in case that happened. If it did, she would have to inject all around the leak to prevent damage.

I knew right away that the red chemical was weird. My skin felt funny, a little warmer, but also like it was drying out. It felt funky.

Finally, the red stuff was all the way in. I could tell it was going to be my least favorite part of the battle. The nurse plugged in the third infusion, an IV drip. It wasn't red. That was good enough for me. The third one was easy, so I sat back and relaxed.

After the infusion was finished, Lynda collected me. She'd gotten the prednisone the doctors had prescribed and Compazine, antinausea pills. I checked the label on the prednisone: one hundred milligrams. I couldn't believe my eyes. One hundred milligrams.

When I'd taken it for a knee injury, they'd prescribed five milligrams. And even then I had to do step-down doses each day, four, three, two, one, a half, then done. But this was a much more serious physical issue. It was a hundred milligrams for five days, then stop cold turkey. So I swallowed the first prednisone down. I felt OK, so I left the antinausea meds in the bottle.

When we got home, Lynda made us something to eat and I still felt pretty good so I ignored taking the antinausea medicine. Judy had warned me about skipping drugs for the bone marrow aspiration, but I still thought I could pick and choose how I medicated the chemo side effects.

❧ ❧ ❧

SIX HOURS LATER, I lay down, hoping I could get to sleep. I felt pretty good, felt pretty energized. But there was no way I could sleep. I was on the ceiling. *Great*, I thought, *I'm wired from the prednisone.* I tried to sleep but no luck. I was lit. I stayed buzzed and alert most of the way through the night. Then about 4:00 A.M., I felt the first wave of nausea.

Thirty minutes later, I was nothing but a tsunami of dry heaves. By 6:00 A.M. Lynda called the hospital twenty-four-hour hotline and, best of luck, Dr. J. answered. He told her to get me to the emergency room right away. I had been on the bathroom floor for a couple of hours by then. She packed me in the car and we went to the ER. It was a cloudy Sunday morning. I was trying to look out the window to calm down, but next thing I knew the clouds made me want to puke. I closed my eyes because it was clear to me that everything made me want to puke.

The ER staff got me squared away in a cubicle, then ran antinausea meds into my IV. The drugs stopped the nausea. I never blew off taking the antinausea meds again.

It was one of my first lessons in paying attention. I would soon start to apply it to the visualizations Kaye had recommended. She had told me to do them before I started treatment, but I had done them halfheartedly and not very often, more to please Kaye than out of faith. The word *visualizations* bugged me. Other cancer patients had assured me I wasn't unusual in not wanting to fight all the time. We all got tired of our whole lives being about beating the enemy.

I started to see that I needed to use every weapon available to me. I finally understood I had only one shot, and if I lost I was dead. And I didn't want to die.

My focus kicked in, and a few days after I got back from the FR, I started doing the visualizations as if they meant something to me. Lynda was a great help in getting me the space to do them. She made sure that I had complete quiet and that Mr. Timms was confined to his quarters. He thought we were crazy, and that climbing me as though I was his tree could only enhance the healing.

I'd close the door to our bedroom, take off my shoes, and sit on the bed with both feet on the floor. Then I'd take a few slow breaths.

Once my pulse had slowed, I'd shake my head, then breathe deep and feel my scalp and bones relax—then my neck, my chest, hands, arms, belly—down to my toes.

Once I was nice and mellow, I'd bring in my ally to fight the cancer—Freedom. I'd known right from the beginning of doing the visualizations that they would be about Freedom and me creating my healing together. She'd dive in from above me and slowly fly throughout my body. I could feel warmth spreading from her flight.

In the next part of the visualization, Lynda and I would

be sitting in one of the exam rooms at the clinic. I was confi-
dent. And I'd feel even better because there was a big window
and I could look out at the sky. Dr. J. would come in with his
smile and give Lynda and me the good news. The cancer was
completely gone.

Then I'd move the visualization forward to being on
the Skagit River with Freedom. We'd be on the spot down
near the water where I had once been alone on that snowy
day. This time it would be raining. The drops were so big we
could walk between them. The sun was shining softly behind
us. It gleamed on the drops so that they looked like melted
crystal. Then, as I would feel Freedom's welcome weight on
my arm, I would see the light on her white head. She would
have come to full maturity; the dark feathers of her child-
hood had molted and been replaced by shining white ones.
We were both far into our mutual future. We had made it.

The nausea also led me to an alternative medicine. I met
with Dr. J. before the second infusion. I asked him about using marijuana to curb the nausea. He offered me a prescrip-
tion if I wanted it.

"No, thanks," I said. "I have plenty of friends who will
help me out." I hadn't smoked pot for years, but I knew there
were people using it effectively to curb nausea.

One of my best friends, Amy, bought me a pipe and a bag
of weed. I kept it for a while, then after the second and third
cycle, the nausea kept getting more and more extreme. The

pills I took for nausea were barely keeping me from puking, and there were days I couldn't move off the bed because the cramping got so bad. I would lie in a fetal position bloated with gas and racked with pain.

Finally, one day right after the third infusion, I crawled into the living room, packed a bowl with weed, and took two or three hits. The pain, cramping, and nausea were immediately gone. I couldn't believe how quickly and completely it worked. I got off the couch and went on about my day.

One of the pills I was taking cost sixty dollars each, and I had to take it three times a day. It was amazing to me what a little plant that grows wild could do that the pharmaceutical companies couldn't. From then on, for the three to four days after each infusion, I smoked a little pot. It helped me get through the toughest times.

Pot couldn't help me with the other side effects, which showed up almost immediately. Even though I'd joked about going bald, I hated the hair loss enough to get my head shaved before it progressed too far. I refused to wake up with a mouthful of hair. And I didn't get my head tattooed.

There were also what might seem like minor irritations, but they could add up. For instance, the treatment center had apple juice boxes and crackers in the lobby for us to eat. It was well intended, maybe as a distraction, maybe for people whose stomachs were still uneasy from chemo. The first time I went in for an infusion I had some crackers and apple juice.

After that, I could never look at apple juice and crackers again, much less eat them. It got so bad I would feel like puking if I even looked at them.

I was able to surf most of the discomforts, but the red solution always made me miserable. It made me need to urinate frequently. I'd be hooked up for the third infusion and I'd immediately have to pee. I'd have to drag the IV pole and the whole contraption to the john. It was an ordeal trying to get that thing in the bathroom, winding past others in the infusion room, and a little humiliating too.

Some of the other patients in the infusion room were a pain in the ass. One woman whined and complained to the point that the nurses closed the curtains on her. There were a lot of long faces. I remember thinking, *These people are bringin' me down.* I suspected they couldn't help themselves. Sometimes I'd notice someone disappear. We would have exchanged nothing more than a hello. Still, I couldn't help thinking a little about where that person had gone, that he or she might not have made it—and I tried not to think about where I refused to go. I would focus on Freedom, Lynda, on Mr. Timms, on living and enjoying life.

Jimmy, the receptionist, was a real comfort and help. His unfailing kindness to me, Lynda, and all the patients never seemed forced. He loved cats, so Lynda and I brought him photos of Sasha and our home cats. We always felt a little more relaxed after he greeted us.

I used anything that worked to keep my spirits up. The oncology staff told me I should avoid going to the wildlife center, but I wouldn't consider it. If I couldn't be with the wild ones, then what was the point of going through this? I even kept doing educational programs with Freedom—the only time during this period that I liked being in a crowd. But sometimes I'd wonder if it might be my last one. Then they wanted me to wear a mask at Sarvey, and I wouldn't do that either. Dr. Jacobs caught on immediately that I needed to spend time at Sarvey—as medicine.

About five days after the third infusion, when I was well into the groove of the chemo ritual, I went up to Sarvey to take Freedom out for a walk. After I worked with Freedom, I wandered around inside just checking things out. A baby raccoon was hiding behind one of the large greens as it was being cleaned. I reached around to get him, and he bit my finger hard enough to draw blood. All I could think was that I hoped the chemo still in my system wouldn't make him feel as sick as I did.

The next day I went to the hospital for a checkup. When the nurse asked what happened to my bandaged finger, I couldn't resist. With all the seriousness called for, I told her, "A raccoon bit me."

She was horrified. "A raccoon!!!!"

"Don't worry," I said, "it was a small raccoon." I held my hands apart about six inches. For some reason that clarification didn't make any difference to her.

My main sources of support came from Lynda; my sister, Jill; my mom; Freedom; and some of my animal friends. Jill was a big help during the chemo. We talked a lot on the phone. She just listened and didn't give advice. She eventually came up and hung out for a long weekend. We listened to music, went out for dinner, and watched a few movies. We kept it simple. We laughed a lot through that visit.

When Lynda needed a break, Mr. Timms was my best therapist. He could make me laugh when no one else could. Even if it involved Lynda as his fall guy. One morning, I heard screaming in the kitchen. I dragged myself out of bed to see Lynda hopping on one bare foot, with Mr. Timms attached to the other. "Do something!" she yelled.

I grabbed him. He chilled. When Mr. Timms had come into sexual maturity, he had realized that his saviorette was a female. Guy squirrels don't have time for girl squirrels unless it's to make baby squirrels. Not that Lynda cared. She loved the little mischief maker. She called him Mr. T., and every day she'd say, "How you be, Mr. T.?" He'd say, "Mmm, mmm," in response. But she wasn't to cuddle him.

Her toast, on the other hand, was most desirable. Lynda ate breakfast in the living room. She'd sit on the couch with her toast and cereal on the chewed-up coffee table. Mr. Timms would race across the table and snatch her toast.

"No!" Lynda would say. Mr. Timms would give her the mad brown eye, slam his tiny front paws on the table, and say, "Erhhh. Erhhh." Lynda would demonstrate her great understanding of behavioral training and give him a scrap of toast. He'd run off with it, take maybe a bite, and let the soggy leftovers land where they would. And, strangely enough, the scene would repeat the next morning.

Lynda and I rewrote the lyrics to the theme song of *Walker, Texas Ranger*, recasting Mr. Timms as Walker, and we would sing it when we watched the show, "When you're in Texas look behind you / Cuz that's where the squirrel's gonna be."

In fact, the mighty squirrel was asleep in his little house. But the truth didn't stop my sister, Jill, from later enrolling Mr. Timms in the Texas Rangers Law Enforcement Association. To this day, I have a card tucked into my favorite picture of him. The card reads: "Tee B. Timms, member Texas Rangers Law Enforcement Association." Tee B. was for one of his nicknames, T-Bone, because he bit to the bone.

On Halloween, he was Mr. Timms, the not-so-wily Snickers thief. There'd be a rustle in the candy bowl on the counter and the next thing I knew there was a gray flash down the hall, with the Snickers mini-bar flapping over his head.

Remodeling? Mr. Timms was on the job. We'd left the sliding door open to the storage closet in Mr. Timms's room. At that time, his cage door was open so he could come and go at will. One day, Lynda and I walked into the room and saw that half the molding on the closet door had been shredded.

We slid the door back and discovered a gigantic and well-made nest of paper towels on the top shelf. Mr. Timms's tiny head popped up out of the nest. "Errh. Errh." He grunted fiercely. We backed carefully away, laughing.

In Mr. Timms's mind, what was yours was his and what was his—was his. However, he was generous in sharing his room. When Freedom was a guest and stayed in Mr. Timms's room, the squirrel never gave her a second look. He was, after all, a squirrel with a mission, and no eagle could get in the way. Mr. Timms didn't see any of the wild ones as danger. He shared a room with owls, hawks, eagles, and falcons and never gave them any mind. He had no reason to. He was Mr. Timms.

I adored him, and he was my comic relief. But more than anyone human or wild, Lynda had dived without hesitation into what we both knew was our journey. She did everything, all the work and research, learning everything she could about alternative treatments that would strengthen me and supplement the healing effects of the chemo. She was the one with the detailed questions for Dr. J. and any other clinicians. She was there for every treatment. More than even all of that, she didn't fall apart in my presence. She did all her crying in private. Her dedication allowed me to focus simply on healing. We were a team. And I was a lucky man to have an ally who often didn't even ask what I needed. Lynda just provided it.

Early on, I thought about the word *ally* and looked it up.

The common definition is obvious—to give support; but the second meaning—to combine resources with another for a common goal—says even more. Sometimes it seemed that Lynda's contributions to my cancer battle were much greater than mine.

AFTER THREE CYCLES OF CHEMO, I went in for another CAT scan. The tumor in my spleen had shrunk dramatically. I thought that I was going to walk through the rest of the treatments. I figured I already had cancer beat.

I did another two cycles of chemo, and Lynda and I once again went back in for the results. This time they weren't what I was expecting—or wanted. The tumor had shrunk but not much. Dr. J. was calm, but what he told me made me anything but calm. I got pissed off and closed down.

I still had another three rounds to go and was starting to wonder if I was going to make it. I couldn't let myself think like that. I forced those thoughts out, but they would keep intruding. Lynda and I left Dr. J.'s office. I didn't really want to talk.

As we walked over the sky bridge back to our car, she took my hand in hers and said, "Don't give up."

I was quiet. I wasn't anywhere near giving up; I was closing down because I was angry at the cancer for not doing what I wanted it to do, which was be gone. And I was distant because I was just plain cranky.

I was also scared in a way I hadn't been before. That was a feeling I didn't like at all, a cold ball of fear in my belly. Being afraid for my life was the real deal. I didn't want to be scared, because in my mind that opened the door to dying. If I was scared, there was a chance the cancer could win, and that could not happen.

Lynda was the rock in our journey; I knew my death would shatter her life. But I also knew she would pick up the pieces and go on like I would want her to do. My family would close ranks and help one another. But I kept thinking that Freedom was in so many ways a helpless wild one. I was her protector.

I turned to Lynda. "I'm not giving up," I said. We didn't have to say more.

On sleepless nights right after an infusion, troubling thoughts would come to me: What if I died? Where would Freedom go? Who would take care of her? What was really hard was that she was still young and our connection was still young, though it was growing every day. I'd start to feel the fears and I'd jam them down. The cancer had happened, and being scared or bitter wasn't going to help me fight the enemy. Our enemy.

And I knew without a doubt that the spirit I called Grand-father would not have me die before Freedom matured and got her white head feathers. It might have sounded like crazy thinking to some people, but it made total sense to me.

In trusting that Freedom and I would have a long time together, I'd found the advantage—call it an edge—that I needed beyond even extraordinary human support. Freedom and I were balanced on that edge.

CHAPTER SEVEN

THERE WAS A PATTERN to the infusions and my body's responses to them. "Responses" is a flat-out euphemism. My body's reaction to the chemo was a flat-out nightmare.

I was having infusions every three weeks. For the first two treatments, I'd feel like near-death the first three or four days. It was as though every cell in my body was sick and I couldn't take anything to kill the pain. Except it wasn't just pain. It felt like my body wasn't my own. That intensity lifted after the first week of the infusions. Weeks 2 and 3 felt almost normal. I figured after the first few, it would get easier. I was precisely wrong. After the second treatment, I began to feel like crap into the second week as well as the first. I realized then that the sickness was going to last a little longer each time. And that was just how it was going to be. There was no way to run away from any of it.

Sometimes I had very disturbing dreams. In those dreams I was fine, no cancer. I'd wake up thinking, *Cool, what a relief, the cancer is just a dream.* I'd go into the bathroom, look in the mirror, and there in front of me I would be, with a bald head. I'd stare back at myself, and reality would come crashing back down.

I felt more trapped than I could ever remember. And then my world opened up. A pair of merciful talons cut through my prison.

One night, feeling bone tired, sick, and scared, I'd managed to fall asleep. Suddenly Freedom came out of nowhere, a tiny speck that grew into a thunderbird. I knew I was dreaming because her wings were perfect, just like they couldn't be in real life.

Even though I was asleep, I was alert to her every move. She soared and banked hard, left and right. Through all her moves, she grabbed and crushed the cancer cells with her talons. Sometimes she would seize the cancer and rip it apart with her massive beak. She never stopped flying and attacking. I would cheer her on. In those dream moments, I felt on top of the world. I had a secret weapon—my winged friend, Freedom.

I lay peacefully in my dream and watched her work out. All the while I was thinking how great that she could fly. I was as ecstatic as I knew she was. Back and forth she went, all throughout my body, and then she flew away, farther and farther until she was only a tiny speck—and then gone. When I awoke, I wasn't sure what was a dream and what was reality. In these dreams, I knew without a doubt that I would survive. On waking I would feel truly secure.

She was my protector. I knew I'd be the same for her. I would die for her. That is not an exaggeration. Sometimes a human being is lucky enough to feel this kind of devotion. I knew I was. I knew I had something powerful on my side.

I knew the Great Spirit, Grandfather, was with me, and Lynda was watching my back, helping me fight. Freedom was my "muscle," fighting alongside me. I do not believe that Freedom was a healer, but she showed me how to look inside myself and find my own strength. And I knew more and more that I was fighting not just for my own life, or for me and Lynda, but for Freedom.

I'd begun to suspect that my fight against cancer was spiritual as well as physical. In some of the darker moments of feeling pain and limitation, I almost found myself up against a wall within myself. I wouldn't give up, but I didn't know where to go next. And I knew there wasn't a pill or shrink or self-help book that could help me find my way. Most of the time, I'd just sit tight with my doubt or fear and ride it out.

A woman at the hospital told me about a woman Buddhist named Pema Chödrön and what Chödrön had said about those moments when there's nowhere to run. "Suffering begins to dissolve when we can question the belief or hope that there's anywhere to hide." I thanked Grandfather that Freedom was both warrior and shelter in those moments there was nowhere to run. I'd slog through the hard days and remember her flashing talons, the fighter-plane maneuvers she made in the air, and the deadly accuracy of her moves.

As I continued to take the infusions and talk with the other patients, I began to understand that Freedom was not the only way to fight cancer. I saw that some of the other patients had a calm that carried them, and I realized that any of us in that battle would do well to go deep inside ourselves, pay attention to the world around us, and begin to trust our own strategy. I knew that the battle plan lay inside.

The doctors could only do so much. The chemo could only do so much. The visualization had its limits. There were no guarantees. But, as time went by, I would have bet that a balance of strong willpower and strong surrender are the perfect medicine. I'd come to understand that surrender doesn't mean giving up. It means giving in—to everything you can learn by fighting.

Fighting a deadly enemy for the sake of someone else is life-affirming. Freedom was my someone else. I wanted to see her mature. I needed to see her mature. I wanted to see her dark head morph into the radiant white beacon of an adult eagle. At a little over two years old, she was already starting to show white on her head, but it wouldn't be pure white for a few more years. That was for me. But I needed to live for her.

I didn't know for sure what had happened to her or what

caused her fall, but her emaciation had told me that she most likely had hatched after a sibling. If she wasn't fast enough to get the food her parents brought, maybe that explained her emaciation. The fall might have been caused by her trying out her wings and slipping over the edge of her nest.

I wondered if ripples of her pain and fear had spread out from her, because when I lay helpless, suffering this visit from cancer, I felt ripples of healing and reassurance coming in from Freedom. I'd think of a pebble falling into a dark pool and ripples of light traveling out in tiny waves—ripples of survival and trust from Freedom to me.

Freedom's dreamtime visits carried me through the chemo months, and eventually I knew she was with me always. I wasn't some off-the-planet visionary, some guy sitting around in a fancy robe hearing spirits' voices. I worked when I had the strength to, rode my bicycle when I could. When I had an appetite, I loved the Mexican restaurant in my town, and I really loved its shrimp fajitas. And, as the infusions progressed, I knew that Freedom accompanied me every step of the way.

I was about ten minutes into the fourth infusion when I hit a wall. I wanted so badly to leave. It took all my strength to not rip the shunt out of my hand. But even as I felt so trapped and furious, I almost could see Freedom's dark-brown eyes looking at me steadfastly, as though she were saying, "I trust you to keep yourself alive."

I had to remember that vision in the months to come.

One of the nastiest side effects of the chemo was that my entire digestive tract shut down about halfway through the regimen. I was so bloated I could barely stand up, and when I lay down, there was no position that was comfortable, not even for a second. The pain was sometimes constant. That would go on for days. Using pot for the pain helped a lot, but I don't think I could have made it through without knowing Freedom was with me.

Sometimes I would almost see her; sometimes I could hear the rustle of her feathers; sometimes I could smell her breath. There was a picture of Freedom on the wall above our couch. She was still a baby, no white on her head. She looked straight out at me. When I was feeling the worst, it would always remind me that as soon as I was a little stronger, I would be able to hold her on my arm and take one of our walks down by the pond. Standing there in the rain with Freedom is one of the most peaceful feelings I have ever known.

Fighting cancer is a full-time job, and I hated giving up so much of my life. There wasn't enough energy for me to focus on anything but the battle. I couldn't go to work on a regular basis. I couldn't enjoy food most of the time. I couldn't take my photographs, write a letter, damn near read the paper without the word *cancer* sticking itself in my face and the chemicals in my body making me feel like an alien. But in the moments that I spent in direct connection with Freedom, there was nothing but easy joy.

The best times were when I'd have enough strength to

visit Freedom and take her for walks. In those moments, the world would disappear, and I'd forget about everything but her and me and the beauty we were walking through.

I'd drive up to see Freedom in my dad's old silver Nissan truck. We'd carried hay in the back, and there was hay actually growing in the corners. I had never cleaned it out. The window between the cab and the back was still smudged with nose prints from my dad's old dog Taylor. After Taylor died, my dad had never washed those prints away. Neither had I.

I loved the air of those summer drives—softly cool and clouds of pollen almost like snowstorms of white reflecting the sun. I smelled the scent of vegetable life, of the persistence of things growing. I'd still feel a little nauseated, but barely noticing it. After feeling so shitty, so mega-shitty, feeling only a little shitty felt really good. Besides, I was out of my living room, out of my house, on the road headed to be with my dear friend.

My visits were all different, but in some ways they had a pattern. I'd pull into Sarvey. I'd hear Freedom calling. She knew the sound of my truck. I'd go back and say, "Hey, how are you? I'll be back in a little while," then I'd head to Kaye's house for the chat we always had. We'd talk about nothing in particular and everything.

I would then go to Freedom's flight, jess her up, and take her through the flight door, the hallway door, and the gate to the back of the compound. Once out, we'd be on top of the little hill on which Sarvey stands. There was an apple tree there. I'd noticed the difference in the scent of the air around the tree, a sweetness, a kind of promise of more. I'd carry Freedom down the dirt drive, across a dirt road, and head up a little hill. The sides of the hill were thick with cedar, fir, and oak. It was like walking through a beautiful 3-D movie, the shadows of all shapes and colors, the path underneath dust-brown with little stones.

As we topped out, the trees were gone and we could see all the way to the green-carpeted mountains to the northeast. Usually it was breezy, and Freedom would spread her wings. When the wind brushed her, her flight instincts would kick in. She'd spread her wings and try to fly. I could see her feathers ruffle in the wind.

After staying out for a half hour or so, Freedom felt like she weighed fifty pounds on my arm. But I loved it. What cancer? Did somebody say cancer? I only knew the breeze and sun and this welcome weight on my arm. I'd look at her beak in the bright sun. It seemed as though it was carved perfectly from obsidian, gleaming black with depths of blue—infinity blue.

By then, I'd feel my energy start to fade, so we'd turn and head back to Sarvey. Freedom's favorite part came right before we got to the center. I'd run on a short paved road with her. She'd spread her wings and face into the wind. She knew she was flying. She was free.

Back at her flight, we'd go through our routine. I'd bring her a turkey leg. She'd eat, then she'd splash into her swimming pool and take a bath. If you had asked me in those moments if I believed I was dying, I would have said something like, "Not today."

Between the seventh and eighth infusions, my white blood cell count had skyrocketed. When I saw Dr. J., he told me that if they gave me the last infusion while my white blood cell count was so high, it would do more harm than good. It could seriously damage my liver. By then I was more than ready for the Spook House ride to end, but only in the way I wanted it to end. I needed to be in control. I was tired of reacting.

I was pissed off again. This roadblock definitely put a crimp in my plans of how this was to finish. I asked Dr. J. what we were going to have to do. He told me if my white blood count got any higher, he would admit me to the hospital. No way was I going there. I had never been in a hospital in that fashion, stuck in a bed while they tried to keep me alive—and I wasn't about to start now. I hoped he wanted to wait and watch my white blood cell count and, when it was low enough, clear me to take the last infusion.

Frustrated and anxious, Lynda and I left the hospital. I had to wait it out. I'm not normally patient—and this wasn't normal. But cancer demands it. You have to pay attention. I had come that far and wasn't going to quit at that point, so I bit the bullet. I waited.

After two more weeks my white count was low enough to proceed. Lynda and I went to see Dr. J.

Dr. J. said, "So Jeff, do you want to take the last treatment?"

"Are you asking me?" I couldn't believe my ears. He was the doctor. I was the patient. He was supposed to be telling me what to do.

"Yes," he said and smiled. "It's your body."

I knew for sure in that moment that Dr. Andrew Jacobs was more than a doctor. He was a healer. He understood that only the patient could affect the deepest levels of healing—and only in a team effort.

"Might as well," I said. "We've come this far."

And it was *we*. It was Lynda and Freedom, Dr. J. and his staff, and me.

The aftereffects of the eighth infusion two weeks later were predictably the worst. And the best—because even though I felt like shit, I could almost believe that the worst was over. And my head looked less like a billiard ball and more like a fuzzy egg.

THE DAYS GREW COLDER. One morning in November after the last treatment, I felt good enough to go jess up Freedom. I would hear in two weeks whether the cancer was gone, but the prospect of seeing my friend smoothed the edges of the fear that was running under the surface.

As I pulled on a jacket, I realized I had automatically grabbed my dad's old winter marine coat. I stopped for a second and felt its sturdy warmth. That coat had taken him to

Korea and back. He would have loved that I wore it to see my friend. When I went out to the truck, I took a second to look at those old dog nose prints.

I got Freedom out of her cedar flight, jessed her up, lifted her high at the gate so she wouldn't bate and crash into the fence, and headed over to the apple tree.

The sky was flat gray in all directions. There was a strong enough breeze that particles of cedar and fir were flying through the air. Inside me, the fear was tamped down pretty good—and it was still there. I sucked in a good breath of air as though I could wash the fear out of my chest and belly. The cold filled my nose and mouth. I tasted winter coming on, and I thought how a forty-degree day in April is so different from a forty-degree day in November.

We headed down the hill on the little dirt road. Freedom spread her wings. I could feel the pebbles and rocks crunch under my old boots. The dust of summer was gone. I wanted to smell that dust again. I took another deep breath and thought how Lynda and I always say at the beginning of autumn, "I can smell the pumpkins in the air."

I carried Freedom up the hill under the heavy canopy of the trees. I kept going. I needed to see out. I needed to forget the feeling of being trapped in four walls—with that sandpaper fear.

We reached the top. I stood quietly. The tall grasses and weeds twisted in the wind. Freedom leaned into the wind, head stretched forward, her eyes even more alive than usual. We both looked out at the low, green-splotched mountains. I imagined the trees on their flanks shaking in the wind.

My skin prickled. I hadn't shaved for a while, and a weird splotchy growth of beard trembled in the breeze. I hated how my skin felt and looked, saggy and gray as though something vital was missing from just underneath it.

I was trying to ignore that feeling when suddenly I felt, not just Freedom on my arm, but my dad's coat warm around me. It gave me comfort, and I knew that wherever he was, he had already taught me how to fight a battle with as much grace as I could manage.

I had learned from a patient at the hospital that when the ancient Mayans played their ferocious ball games, they were said to either play for the jaguar or play for the worm. If you played for the jaguar, you were playing your heart out. If you played for the worm, you were playing with less than your heart. The Mayan ball players were playing for their lives— those playing for the worm would be sacrificed.

On those brief walks, for those always too short moments, I knew for certain that Freedom and I were playing for the jaguar. We were playing for my life and, in many ways, hers.

CHAPTER EIGHT

I KNEW I WASN'T GOING to hear the results of the posttreatment blood work until just after Thanksgiving. Sometimes I was glad about that, sometimes I wasn't. Mostly I hated being in limbo.

Dr. J. had told me if I still had cancer after the eight rounds of chemo, I would have to check into the hospital for a stem cell transplant. There'd be an intravenous catheter in my neck, sterile room, side effects from the preservatives in the stem cells, including more vomiting than I wanted to contemplate. The transplant could save my life, or it could cause stem cell failure, organ damage, blood vessel damage, cataracts, more cancer, and death.

Even if the cancer was gone, there'd be years of testing and more stretches of limbo.

The days between the eighth infusion and hearing The News were some of the longest days of my life. I was so grateful for Freedom, though the moments I spent with her always went too fast.

Finally it was Thanksgiving. Freedom got her turkey leg. I'm sure that hers tasted a lot better than mine. Food tasted like sawdust that day.

The day after Thanksgiving I went in for my blood draw. Then it was the day after the blood work, and then the day after that.

And then it was Monday. Lynda and I drove down to Dr. J.'s. We talked a little. I was hopeful—but there was a tiny current of raw fear in me.

We drove our familiar route through the arboretum. The sun was a silver disk low on the horizon. Skeletal trees stood in the wet gray light. There was a sleepy feeling in the November air, as though the earth was heading for a hard-earned rest.

We turned and drove up an average city street, with little stores, condos, lights, business as usual. I watched people going about their daily lives. I just wanted to be one of them, worrying about what was for dinner. I was tired.

And then we popped up over a rise. There was Puget Sound gleaming below us. I'm a water boy with salt in my veins. Being near the ocean is always medicine.

This natural infusion of salt, moisture, and light carried

me out of the car, across the parking lot, over the sky bridge, and toward the slowest elevator in the world. The elevator doors opened and I could feel my ease going out on an ebbing tide. The elevator creaked upward at prehistoric speed. Lynda and I were quiet, each in our own worlds, though our fingers were linked. Centuries later the elevator doors opened one floor up and a few people got on. We left the Triassic and headed for the Jurassic, another floor up. Finally we arrived at the Modern Era. The door opened, and it was make-it or break-it time.

We moved through the check-in routine. As always, I caught myself watching people. I couldn't not. I swear I could tell who was going to make it and who was dying right in front of my eyes. I wondered who was watching me—and what they saw.

Lynda and I sat in the lobby. One of the nurses walked toward us. The nurse led us into the room where Dr. J. would talk with us. I knew immediately that there was something seriously wrong. There was not one window in the room. In all the visualizations I'd done, I'd been sitting in a room with a window looking out over Seattle when Dr. J. told me the cancer was gone.

There was no window. Panic bubbled just under my skin. The nurse checked my blood pressure. It was through the roof. The nurse looked puzzled, but she had no way of knowing about the lack of a window—which I knew was going to kill me—or the absolute fear that swept over me for that instant. All the way through this cancer ride, that by far was the most frightening moment.

Lynda didn't know what was going through my mind at that moment. How could she know? I'd never mentioned the window. I didn't want to scare her. So I resigned myself to what was coming down the road.

When Dr. J. came in, he was smiling. I wondered if I'd misjudged him and he was a closet sadist. He did his usual greeting routine—both to check on how I was and put us at ease. He opened the chart and said something. I couldn't process the words immediately.

Then I heard, "You get to live. There are no signs of cancer anywhere."

I think Lynda didn't realize what he had said. She had a slightly confused look, until I said, "It's gone." We hugged. Then it was Dr. J.'s turn for a hug.

He told me about an experimental drug, Rituxan, with which I would do four treatments, because it seemed to anchor the remission. I signed on right away. Then he told me I would have to come in for checkups every three months for the next two years. I was so high on the good news that I figured those wouldn't be any more trouble than mosquito bites.

We left. Lynda walked up to Jimmy and said, "We won the lotto." His smile was bigger than ever.

I dropped Lynda off at home. We'd been through an eight-month-long roller-coaster ride. I knew she needed a little time alone with herself and the cats. She knew I needed to take Freedom for a walk.

I headed straight up to Sarvey. As I drove through the silver mist and cold, I felt the weight of the journey lift. There was room in me for pure joy. Although I knew there was another course of chemo ahead, I was traveling a familiar road that seemed more beautiful than ever. And I was not only heading up to see my feathered friend, I was heading into three months of knowing I was healthy. Knowing, not just believing.

I turned up the dirt drive to Sarvey, parked, got out, and stopped to tell Kaye the news. "I knew it," she said. We chatted a little longer, then I left. I looked back and saw her sitting in her window aerie, the spot from which she could survey her domain.

I grabbed Freedom's leash and jesses. She hopped onto my arm and I jessed her up. Each action, each moment, her scent, her dark eyes, my heart pounding in my healthy veins—all of it was perfect. There was just her and me—nobody outside—and the good cold air I love so much. We went out through the usual doors and gates with ease.

We stood outside looking out over the valley. All the muted colors were as vivid to me as if the sun shone on a summer day—the greens, the browns, the soft grays and silvers. Freedom was on my right arm facing me. I felt Freedom's injured left wing drape over my right shoulder.

Then I realized she had brought her right wing clear around to touch me in the middle of my back. I could feel the tips of her primaries pressed into my body. I could feel the bones in her wing on my shoulder. I looked down and saw the ends of her feathers as though I wore a dark eagle cape. In that instant I felt my friend enfolding me in a full embrace.

I leaned back. She had never done this before. We looked straight at each other. We both leaned forward. She gently touched her beak to my nose. I don't know how long we stood that way. The world was gone. It was just Freedom and me.

CHAPTER NINE

I WAS DRIVING TO WORK via the Highway 520 floating bridge across Lake Washington into Seattle. The Cascades were behind me in the east. I could see the Olympics to the west. Mount Rainer in the south and Mount Baker in the north caught the morning light. Those two mountains are volcanoes and are part of "the Pacific Ring of Fire"—a string of volcanoes that circle the northern and southern Pacific Ocean. When it's not cloudy, the sight is spectacular.

I was past the halfway point on the bridge when out of the corner of my eye I spotted a white head skimming no more than a foot above the water. It was a mature bald eagle. In between us were logjams lined with cormorants—big black fish-eating birds with nasty hooked beaks. They were facing in my direction with their wings held wide to dry.

The eagle was in full stealth mode as he closed in on the cormorants. He flew parallel to them. Cruising with intermittent flicks of his wings as he picked up speed, he made a casual left turn as he went behind the birds. They didn't seem to perceive the threat. He started to accelerate by pumping his wings—then he was in the middle of them. There was an explosion of feathers. He emerged with something in his talons and headed straight for me.

I was coming out of my skin, motoring sixty miles per hour across the bridge. I looked around. All the other drivers had their heads down or were staring straight ahead. I looked back at the eagle. He was headed my way with what had to be his cormorant breakfast in his talons.

The eagle came right at me toward the right side of the truck, no more than ten feet above me. I looked at him. He cocked his head and looked straight at me. We were so close I could see his pupils.

Then he was gone. I was on fire. I looked around again at the other drivers. I wanted to wave at them and share with someone the miracle I had just seen—but nobody was looking.

AT THE END OF February 2001, I walked into the center and found we had a special new patient. A Sarvey volunteer had picked up a tiny black bear cub weighing no more than seven pounds from a person who thought he was being a Good Samaritan.

The man had been walking with his dogs in the woods on the Olympic Peninsula. He'd let the dogs run free. They came upon a black bear mom and two cubs. The mother chased the dogs off and made her getaway with one cub. The man thought that the mom had left the second baby behind, so he picked up the cub and took her home.

I looked at the lonely cub. I knew what had really happened. Black bear mothers are ferocious defenders of their young, and if they leave a cub behind, they will come back to the area for a week or more looking for it. If the man had left the bear cub alone, the mother would have come back and rescued it. That hadn't happened, so we had a baby bear at Sarvey—and we only had one.

In bear rehab, it's much easier to have two or more cubs. Like most species, they learn from play and have an emotional need for their own kind. One lone cub is much harder for humans to raise and release.

The little bear began her temporary life at Sarvey inside. She stayed inside from late winter to spring. She had formula for bears in a bottle, and then bowls of bear mush, fruit, honey, bugs, fish, and more. We tried to keep the human contact to a minimum to avoid her becoming habituated, but it was hard while she was inside. If we'd isolated the cub, it could have emotionally damaged her. She needed one "mom," and that was going to have to be a person.

One afternoon when I was feeding the cub, Sue Mc-Gowan, Sarvey's clinic director, suggested that I take the role of mom. The bear cub would identify with me and treat me as a parent. It was great timing for me, because I could get to Sarvey more than once most weeks. I was still weak from the chemo, so I wasn't getting to work more than three or four days a week, but a few hours working at Sarvey made me feel good.

I knew that unlike a real mom, I would eventually have to let go of the cub. If we were careful not to let her get habituated to people, she would revert to her wild instinct and immediately become fearful of all people when the first cold snap hit. She would become afraid and run away—easily spooked and a little standoffish, even with me—which would be exactly what a responsible rehabber wants. At that point, I would stop all play and interaction with her.

Still, I jumped at the chance to be her mom. As the cub continued to grow, her personality began to surface. She was a rambunctious child. She threw tantrums, screaming and destroying whatever she could get her paws on. Mother bears teach and discipline their young, but this cub had no one but the volunteers at Sarvey, and me in particular, to help her along. We would do the best we could

Like most bear cubs she was really curious. She'd climb up on top of the six-foot-high squirrel pens to explore—and then go to sleep. We'd put a food bowl on the floor for her and when she was eating out of it, she would sometimes bite whoever walked by. She'd bite at people's ankles and nipped a volunteer bent over to clean a cage on the butt.

She liked to eat up on top of the unoccupied squirrel cage. Squirrels were just starting to arrive then and were too small to be in the larger squirrel cages. I gave her a bowl of food one day and she swatted it so hard it went flying across the room. Blueberries, honey, applesauce, and dog food splattered all over the wall and other cages. From then on all her bowls stayed on the floor. About that time we named her Angelica; another volunteer had described the wild little girl named Angelica in *Rugrats*. The name suited her well.

After the weather started to warm and she was big enough, we moved Angelica outside into the heavy chain-link mammal pen. This pen was designed for full-sized black bears, though other large mammals were sometimes housed in it. It dwarfed Angelica, but she had two tubs of water and room to run and play—room to climb on huge logs and throw things.

The large mammal pen is divided into thirds by guillotine doors, so I could close Angelica off in part of the cage and clean where I needed to without her being in with me. I kept two-thirds of the cage open for her to play in while I cleaned. If she was in with me, the only thing that got done was playing.

As Angelica's mom, I would clean and feed her and give her deworming medicine. Young bears usually come in with

Kaye with a one-year-old bald eagle, 2005. (*Ken Lubas*)

Crazy Bob is surprised by a raccoon that escaped his carrier, 1996

Kaye with a brown pelican, later released in Southern California, 1998.

Sarvey grounds, 2009. *Left to right:* beaver condo and pool, golden eagle flight, bald eagle flight, hunting flight, and falcon flight.

Freedom, age five months, takes her first bath in September 1998.

Jeff and Freedom, Whidbey Island, December 2000, just after Jeff was pronounced cancer free. *(Ken Lubas)*

Jeff during cancer treatment in August 2000, holding a golden eagle before its release. *(Kestrel Skyhawk)*

Sasha, age twelve years, 1997.

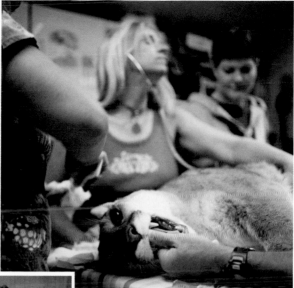

Sue examining Sasha during her annual checkup, 2001. *(Annie Marie Musselman)*

Mr Timms celebrates the holidays, 1998.

Jeff and Angelica, nose to nose. *(Ken Lubas)*

Jeff and seven-month-old Angelica, summer 2000. *(Ken Lubas)*

A snowy owl, native to the Arctic, during rehabilitation. She is now on the Sarvey education team. *(Annie Marie Musselman)*

Leslie holding a fawn, spring 2001. *(Annie Marie Musselman)*

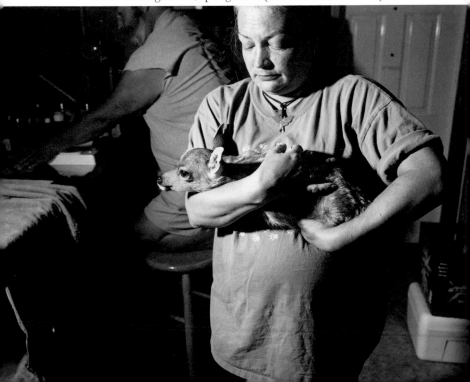

An injured golden eagle being treated for shock, 2008.
(Annie Marie Musselman)

A young bear in a coma following a 2007 traffic accident. He was rehabilitated and released in spring 2008.

Otters in an incubator, spring 2009.

The release of a golden eagle in the Skagit Valley, 2000. *(Kestrel Skyhawk)*

Jeff and Freedom, March 2008. *(Ken Lubas)*

Corky and Pumpkin at Sarvey, May 2003.

Corky, after her relocation to Texas, paw up to greet Jeff, June 2004.

The birds always lead the Grand Entry at the River Fest powwow,
August 2008. *(Leslie Henry)*

Sarvey volunteer Kevin
talking to a golden
eagle at River Fest,
August 2008.

With Sarvey volunteers and education birds,
August 2008. *(Noel Nic' Ferguson)*

Freedom and Jeff on Martha
Norwalk's *Animal World*
show, broadcast on KKNW
AM 1150 in Seattle, 2008.

Freedom and Jeff, 2008. *(Ken Lubas)*

worms. I gave her a shot of ivermectin once a month for the parasites. "Come here, my little pincushion," I'd say in a deep voice. Tiny Angelica would saunter over, doing her slightly bowlegged "big bear" strut. I'd put her head between my knees, lean over her, and stick her in the rump with the syringe. It was all over in five seconds, and it was a game for Angelica. She never once refused to get her shot.

Angelica and I played for hours. I taught her how to stand up and be a big bear. I would get on my knees and raise my arms over my head. I'd wave them around and tell her "Big bear, big bear." Angelica would stand up on her back legs with her arms over head and mimic me while slowly walking toward me. Then she would grab me in hug, and we'd roll to the ground. She was like a teddy bear come to life.

We wrestled every time I was with her. We'd go nose to nose. One of the ways bears identify one another is by sniffing each other's breath. I'd roll Angelica over on her back and rub her belly. She'd jump up and scamper away, running for all she was worth. Then she'd sneak around one of the big logs in her enclosure and get me. I'd see her peeking around a log that was four to five times her size. If I looked and made any move, she'd run the other way and hide behind other logs, then sneak out and creep up on me. I would let her track me and win.

When she got really excited and started running, she'd pin her ears back. Her body would scrunch up as she got a full head of steam. She would set a blistering pace, running up the sides; then she'd abruptly turn and run horizontally along the chain link, down the other side, and do it all over again.

We'd tumble through the pen, and since it was covered in straw, we both rolled in that. She would try to grab me with her front paws, and I would grab her back legs and pull her toward me so she couldn't, egging her on. All this was designed so she could have a chance at a normal life by learning some ordinary things bears do, as well as have fun to keep her mind healthy.

She learned to scratch her back on the logs and tree stumps in her pen. Sometimes she would creep up on top of a log as if she was stalking me. I would look away but sneak a peek, and then I'd spin around just before she was ready to pounce. She would stand up like a big bear waving her arms. "You got me," I'd say. Other times I'd spin around and grab her, lift her high in the air then lower her back down onto her back, and tickle her under her arms. When we were roughhousing, she'd grunt, groan, smile, bite, and laugh as only a bear can. I would bring her eggs as a treat. I started by holding one in the palm of my hand. She looked at it, then up at me. I cracked it open in my palm. She slurped it up and wanted more. Eventually she learned to crack the egg but just hard enough so that it wasn't crushed.

Just like Freedom, Angelica knew the sound of my truck. As soon as I got to the top of the hill I could hear Freedom calling to me and Angelica howling for me. The sound of a baby bear crying for its mother is heart wrenching; you could hear her all over the compound. I would have to make a mad dash out back, open the door that leads to the large mammal pen, and run down the hall. She'd start jumping back forth, climbing the door, and crying for me to come play.

With Freedom calling and Angelica crying, I'd tell Angelica that I would be back and we'd have big fun, but first I had to take care of someone else. "I won't forget you," I'd say. "Just hold tight." Initially, she didn't understand, but she soon caught on. Nonetheless, I always had to go and reassure her first.

I would go to Freedom's flight and jess her up for our walk. I had to do Freedom first because in most of our play, Angelica and I were in her straw bedding. Wet straw can be lethal for raptors. They can get a respiratory disease called aspergillosis, which comes in part from decaying vegetation. Asper is a fungal infection of the airways, caused by the species *Aspergillus fumigatus*. If a bird gets asper, it will cough and wheeze and have difficulty breathing. It can even throw up smelly black fluid as the disease progresses. Once a bird gets asper, it's hard to treat and sometimes even if the bird survives, the bird is never the same.

As the summer sped by, I knew our time together, me teaching Angelica and us playing, was coming to an end. When fall came and the first cold snap hit, playtime ended.

The next time I saw Angelica, she ran from me and hid. She was afraid of me now. I could still feed her and clean her pen, but that was it. I'd been prepared for this, and knowing she was reacting just as she should, just like a bear who will be released, made it easier for me to accept our new limitations. It was sad, but joyful at the same time. Our time was over. Her bear life was starting again.

Angelica wintered over at Sarvey. Bears who live in the low lands don't really go into hibernation; it really isn't cold long enough in the foothills. Even the mountain bears come out of hibernation in the middle of winter and forage. Angelica's metabolism slowed down as she ate less and slept more as winter passed. I knew when spring rolled in she would be leaving us. We wouldn't keep her for another year, and she was old enough to survive on her own. She was exhibiting all the signs a bear should—foraging under logs and rocks, looking for grubs and insects. She had a natural fear of humans, and she knew how to fish. Early on I put live fish in her big trough and I hid so I could see what she would do. Sure enough, after I watched for a while, in she went and came out with a trout. My heart sang, knowing her instinct was strong.

Spring arrived and she started eating a lot more. Her body was filling out and her ebony coat glistened like black diamonds when shards of sunlight streamed through her enclosure. Angelica was a stunning bear.

Spring spreads slowly up the mountain. Bears eat some of just about everything—fish, berries, grubs, leafy plants, sometimes a deer or elk carcass. Toward the end of May, when the weather warmed, the berries in the mountains were ripening, the forest was alive with new spring shoots, the streams were flush with fish—everything a young bear would need to survive—we made arrangements to release her far back in the Cascades. We always release in an area with food and water to give a wild one the best possible chance.

The day came for Angelica's release. It was bittersweet for me. I had gotten too close to her. I had never done that with another wild one who had a chance of release. I had worried so much about how she would rehab as an only bear that I hadn't thought how it would hurt to let her go and have no way of knowing her fate.

We sedated her so we could put her in the bear "tube," a steel enclosure with bars and a sliding door at one end. It was designed to be towed behind a truck while the bear inside was sedated. Since we weren't going a long distance, we administered drugs to help her slowly wake up on the way to her new mountain.

Once we arrived at the release spot, we made sure she was awake and ready to begin her life as a wild bear. We all stood back as the door of the tube was opened and Angelica ambled out, taking her time to look around. Once she decided it was OK, off she went toward the forest. Then she stopped, and turned around one last time, looking back at me. Then she was gone. There was a running stream not far away and we left her some food just in case she came back. My heart was heavy and happy at the same time. She had come to us as an eight-pound cub, and now she was heading back into the wild as a big girl, about 120 pounds, not nearly her ultimate adult size, but ready to live on her own.

It was the right thing for her, I knew that. The way she changed toward me after the cold snap was a potent reminder that the instinct to be wild is incredibly strong. I could never deny her going home.

Angelica was smart and I hope she was a survivor, but that well-meaning dog owner had put her at a tremendous disadvantage. We did our best teaching her what she would have learned from her mother, but we don't know everything that a mother bear teaches a cub, or how. And as a young cub without a real mother, she never learned how to *be a mother*. If she survived and had her own cubs, could she—would she— mother them? The next generation might suffer for her loss. Though I believe we did everything possible to bring her up as well as a mother bear would have, there was no way for me to be a bear.

I would always miss her, and I would never know if she survived. My ride home was very long that day.

❧ ❧ ❧

A GOOD FRIEND TOOK a picture of Freedom and me in early August 2001. When she gave me the picture a few weeks later, I thought of the last scene in my visualizations and how it had come true. My hair was almost grown out, and Freedom's head feathers were about 70 percent white. When eagles come into maturity, their brown head feathers gradually molt and are replaced with the brilliant white feathers you see in adults. I laughed, looking at the picture. Whether we had both matured or not was a moot question.

I went through the Rituxan infusions to anchor the cancer's remission with no side effects. After that, Dr. J. told me I had to come in every three months for checkups. *Welcome to living your life in pieces*, I thought. Every now and then in those first few months, I'd suddenly discover a bump or a pain somewhere on my body. I'd flash, "What's that?" and then I'd just make myself ignore it.

Dr. J. had also told Lynda and me that five years down the road was the more-or-less magic crossroads. I would have exactly the same odds of getting cancer as someone who'd never had it. When a year went by and I was still in the clear, I figured I'd go ahead and count on having a future. There wasn't anything to do but live. So I did just that.

One of the most frustrating things about chemo had been not feeling strong enough to go with Crazy Bob on ambulance runs. I went back out with him as soon as I could. He'd rescued a smart little border collie and named her Billie. She could find a raccoon before we could, and she'd wait patiently till we showed up.

And then Bob warned me that his own health might soon jeopardize our ambulance calls. He had a chronic disease that made him tired all the time. I didn't ask for details. I knew Bob was a private guy. But from then on I knew every run was a gift—like the night when Bob got a call about eleven and immediately called me. The caller had said there was something weird going on with two eagles—one eagle had another trapped upside down in a tree. "I'll be out front waiting," I said.

Bob picked me up, and we drove out to an upscale suburb on the east side. We pulled up to a complete circus—neighbors, cops, the news, and curious passersby. There was a huge blue cherry picker and two spotlights. Car headlights were trained on the circus. There was blinding glare and black shadows—it looked like the ending of *Apocalypse Now*.

I looked up at the eagles and laughed. "Bob," I said, "I know exactly what's going on." He grinned. He knew too.

The mighty leaders of the rescue operation came over to us. "You don't need all this stuff," I said. "You don't need to do anything. The birds aren't stuck. It's a territory dispute."

There was dead silence. I could tell they didn't believe me. The mighty rescue had to go as planned. They started up the cherry picker and moved it toward the tree. The eagles immediately decided they needed to be elsewhere, let go of each other, and took off. Bob and I quietly climbed back in the ambulance and left. We laughed almost all the way home.

FREEDOM AND I CONTINUED to do educational presentations—for schools, some at-risk kids, festivals, and powwows. Her personality in public performances went diva. She played to the crowd, and she loved the attention. She could always hear the difference between when I opened her big carrier to get out my gloves and jesses and when I was setting up the carrier to go to a program. When we were heading out to a performance, I took the front off the carrier, put in newspapers, set her perch back in, and set the carrier right at the back of the truck. The sequence of sounds is completely different from just getting the jesses and gloves.

When she heard her on-the-road signals, Freedom would be raring to go. She'd let me catch her and jess her up without any games. We would walk the slope to the truck. When we were about ten feet away, she'd often launch off my arm toward the carrier. I'd catch her and hold her, because if I put her back on my arm, she'd just launch again. I also had to be holding her as I put her in the carrier, because if I didn't, she was liable in her enthusiasm to slam into the truck or the carrier.

When we were in front of the audience, I'd take her out

of the carrier and her charm meter would go to eleven. She
was beyond mellow. She'd be checking everything out. She'd
burst into baby talk with me and smooch a little. More than
once, we were accused of having a secret romance. I always
took the Fifth Amendment.

Freedom could be a trickster, though. It makes sense,
because the First Nations up here hold the trickster in high
regard. They usually are referring to the raven, but in one of
Freedom's wilder moments, the trickster was an eagle.

On a fine October day in 2001, Freedom, Kaye, a few
other Sarvey folks, and I took the raptor team to the Salmon
Days Festival. The event was held in downtown Issaquah at
the salmon hatchery. We brought a barn owl, a great horned
owl, a peregrine falcon, a red-tailed hawk, and Her Highness.

The festival was a great opportunity to educate people
about habitat conservation and protecting the wild ones. We
spent two days answering questions and showing off the
stars of the show—the birds. Since this festival brought in
more than 150,000 visitors over a weekend, this was one of
Freedom's first big tests meeting the public.

The day was warm for October. We were set up under a
big tent, in part to keep the birds cool. I took Freedom out into
the crowd now and then to meet people as they walked by.

As the day went on we were misting the birds with water
bottles to help them cool off—all except Freedom. She eyed
the water bottle with a great deal of suspicion. She knew it

was a nasty little creature that was going to get her if she didn't keep an eye on it.

By midafternoon I could see that Freedom was getting warm. I thought I'd give the water bottle a try. I made sure it was set to a fine mist because I had seen how nervous she had been watching others get misted. I slowly approached her. "It's OK," I said. "This will help you stay cool." I lifted the water bottle and pulled the trigger.

Freedom exploded into midair over my head. She leaped the full length of the eight-and-a-half-foot leather leash. Anybody watching would have thought I was trying to kill her.

She landed on the plastic lawn chair in front of me and sent that flying into the table that had all our pamphlets and sign-up sheets on it. Everything went sailing. She flapped her massive wings and managed to knock over a couple more chairs. All the others birds bated in alarm.

I picked her up on the glove. She was breathing hard. I could hear her heart pounding. *Great*, I thought, *now she's really hot and upset*. It took her a good thirty minutes to finally calm down. She didn't let that water bottle out of her sight the rest of the day—and I never brought it within range again.

That night I took Freedom home with me. We had to be back at the festival on Sunday and Issaquah was closer to my home than Sarvey. The good folks from the hatchery had given Freedom a big king salmon for dinner. Since it was much too large for her to eat, I'd have to cut it up later at home.

Back at my house, I brought her inside for dinner. I took her into the laundry room by the back door, set her perch on her carrier, and put her on it. Then I cut a good-sized chunk off the salmon. The fish was prime quality, pink and fresh. I held it out for Freedom and she snatched it from my hand. *Good*, I thought, *she's hungry. She might forgive me.* I closed the door to the laundry room so she could have privacy. Lynda and I watched a little TV, but I kept one ear tuned to Freedom, listening for anything out of the ordinary.

No sounds came from the laundry room. After about twenty minutes, I went and checked on her. She had been an industrious bird. Instead of eating the salmon, she had redecorated the laundry room with it. There were little pieces of salmon all over the washer and dryer, the walls, the ceiling, and the floor. The room looked like it had the measles. Astounded, I called Lynda in. We could only laugh. Freedom stood tall on her perch, smiling as only an eagle can. I figured she was done, so I put her in her carrier, and Lynda set to cleaning up.

In the morning I took Freedom out back so she could stretch her wings and get some fresh air. We stood in the yard, and in a few minutes the crows started arriving. At first it was just a few, then a few more. Pretty soon they were coming from every direction, circling over us, filling the air with their raucous caws. More and more came, until the sky was a black whirlpool of crows all telling the eagle to go away. Freedom was oblivious to them; she was enjoying her morning. They upped the volume. She didn't even notice.

Freedom visited with us again a few weeks later. Before we fed her, we taped garbage bags all over the place. Of course she was perfectly behaved, not one speck of salmon anywhere other than on her perch. I figured I was going to have to learn to appreciate eagle wit.

The more I worked in public with Freedom, the more I learned to play with the audiences' heads—especially with kids. When she and I had first worked with an audience, she'd be a little excited. Excited raptors poop. So I would have put a tarp down. And when the kids predictably went "Euuuuuuuuuuuuu!!!" I would take a step closer, swing Freedom out a little bit, and say, "I can aim her!" They'd calm right down.

One time, Freedom, Kaye, Mellow Yellow—Kaye's red-tailed hawk—and I were at an elementary school in Edmonds, Washington. I had Freedom on my glove as Kaye was speaking to the class. "What is another name for a bird of prey?" she asked.

An attentive young man sitting in the back row raised his hand. Kaye called on him and he answered confidently, "Backbone."

There was a moment of silence. All the adults looked at one another, mouthing the words, "Bird of prey? Backbone? Whhhaaat?"

Then it dawned on us; he thought Kaye had said, "What is another name for vertebrae?" The teachers had to turn away as they smothered their laughs. I buried my face in Freedom's chest to keep from laughing. Kaye turned back to me, eyes wide, biting her lip, and with all her considerable kindness and poise, turned back to the eager young man and said with a straight face, "No, honey, not vertebrae, bird of prey."

"Oh," is all he said.

Another time, I was scheduled to give an eagle program at an elementary school in Arlington, Washington. It was late April and a beautiful day, so we decided to have our show outside. The kids were sitting in the stands by the athletic field. As I spoke about Freedom and her kind, the kids shot their hands up. I could barely finish a sentence. I knew what to do.

I abandoned my talk and opened the time up for questions. We were about twenty minutes into the program when I called on a young lady in the back row. She asked the most innocent question, "When is Freedom's birthday?"

Without thinking I said, "Today is her birthday." In this latitude, eagles are born April to May and it was mid-May, so I figured it might be the truth. Before I could say anything more, this same bright girl asked, "Can we sing 'Happy Birthday' to Freedom?"

"Sure we can, right after we finish with the questions." We continued on for another ten minutes until it was almost time for them to go back into their classrooms. "Do you still want to sing 'Happy Birthday' to Freedom?" I asked.

They all shouted, "Yes!"

"OK," I said, "on the count of three, we will all sing together."

At this point the young lady in the back row jumped clear out of her seat, and with all the authority of an experienced bandleader shouted, "ONE! TWO! THREE!" Every kid joined in.

Halfway through the song, I shouted, "Louder!!"

They got louder and louder. They all were screaming at the top of their lungs, "Happy birthday, dear Freedom / Happy birthday to you." Freedom, queenly as always, soaked up the limelight.

Right at the end of the song, the same young lady who started this whole thing jumped straight into the air, pumped her fist, and let loose an ear-splitting "Wooooooo!!!"

Not all my experiences with kids and Freedom were as fun-filled and raucous. For example, once I had finished a presentation at an elementary school and had just put Freedom in her carrier. The big kids were leaving first. A little boy, maybe six or seven, came up to me and stood staring at me in silence.

I knelt down so we could be eye-to-eye and I said, "Hi, do you have a question?"

Without a word, he flung his arms around my neck, buried his head in my shoulder, and hung on. I hugged him back. I didn't know how long we were like that. I finally noticed his teacher standing behind him.

"OK," she said gently. "We've got to go now."

The little boy let go of me and followed her to the door. He never spoke a word.

Another time about a year later, Freedom and I were up on the Skagit River at an annual Eagle Festival. It was Sunday morning, before the second day of the festival. I always liked to take her out for a walk and a wing stretch after we had spent the night.

It was a perfect February morning for an eagle and a guy who loves winter weather—gray clouds overhead, damp air, the feeling that Nature was hunkered down for the long haul. I carried Freedom down a dirt trail into a big pasture overlooking the Skagit.

We stood in the soft cold morning. There were a few eagles in the tree line, doing their *chak chak* talk. The air was so crystalline, we could hear every nuance.

I turned around to head back up the hill. Four people, a mom and dad and two kids, were coming down the trail toward me. I was in the peacefulness of Freedom, the other eagles, the river, and me. *No*, I thought, *I don't want to talk to anybody right now.*

The people walked directly toward me. There was no other way to go back to where we were staying. The mom waved. They all stopped a good respectable distance away. Then the dad came over to me and said, "I hope we're not bothering you. We saw you down here and hoped we could speak with you.

"We're from out of state and we came to the Eagle Festival with the help of the Make-A-Wish Foundation. Our youngest boy is very sick. And he loves eagles. Could he come up and meet Freedom?"

They came up close. The little boy's eyes and his grin were huge. "Do you want your picture taken with her?" I asked. He bobbed his head. He was too excited to say, "Yes!"

I knelt down so he could stand right next to Freedom. I'll never forget the expression on his face—the joy, the wonder, the amazement. Dad took the picture. They lingered a few minutes more, and then we all headed back up the dirt trail.

I got to teach a lot, not just about Freedom, but also about the greater natural world outside the classroom—and how to show respect for the wild ones. Freedom and I were giving our talk another time at an elementary school during the week before school let out.

There was one kid who was smart-mouthing every chance he got. His teacher was oblivious. I glanced at her. She ignored me. Finally, I said, "The next smart-mouth question from you, I'm packing this bird up and leaving."

He shut up. Later, as I was packing up our gear to leave, he wanted to come and help me. "No way," I said. "You didn't give this bird and me any respect. You don't get to be the big man now."

One of the quiet kids came up and asked if he could help. "You bet," I said.

Of all the presentations I gave for kids, one stands out the strongest. Freedom and I especially liked programs with at-risk kids. In a program that would repeat once a year for several years until funding cuts took it away, we visited with inner-city kids at a little campsite in the woods in the summer. These kids came from having nothing. An organization brought them in for free and gave them three days of heaven on earth. Most of these kids hadn't seen any wild ones at all, much less an eagle. Freedom was their first exposure to one.

Our first visit, I took Freedom out of her carrier and set her on my arm. I could sense she was in full diva mode. In one motion I raised both arms to the sky. In flawless synchrony, Freedom moved to my fist and, at the height of the move, spread her wings as wide as they would go. The kids rose to their feet and cheered.

At the end of our presentation, the kids sang us a song. One of the teachers played guitar. The children's voices were soft. The song was about an eagle getting shot. The lyrics were the eagle's thoughts as it lay dying on a cliff, wondering why any human would have done that.

Freedom loved powwows. I did too, especially the Stilliguamish Festival of the River. They had great fry bread, Indian tacos, and a salmon bake—but that was just the food. This festival was held in a gorgeous emerald green valley with the wild Stillaguamish River running through it. The valley seemed to contain the deepest sense of the spirit of not just the eagles, but of the First Nations who are their neighbors, and the rivers and mountains that are home to both eagle and human.

It was an even greater gift to take Freedom to this festival the summer of 2001, after I had been declared in remission. Everything seemed more intense than it had before—the wood-smoke perfume of the cooking, the brilliant colors of the dance regalia, the almost blinding blue of the sky.

It was the first time that Freedom and I took part in Grand Entry since the cancer. As I mentioned earlier, Grand Entry is the formal procession that opens the powwow. The drums always pounded right into my heart and called us in.

Freedom and I stepped out right behind the leader of the procession. He was in full regalia—beaded vest and leggings, feather bustle, feathers in his black hair, bells on his ankles and wrists. He carried a long wooden staff with an eagle head on one end and an eagle claw on the other. He set the pace in time with the drums. I took my first step forward into the procession.

I thought of the dignity of the wild ones. I'd gotten to release a golden eagle the August before when I was in chemo. That had been one of the few gifts in that barren time. Kestrel Skyhawk, the woman who was in charge of the raptors at Sarvey, asked me if I would like to release this eagle. She and I drove the golden eagle to the place it had been rescued—a tall grass prairie.

I took the golden out of the carrier, held her with her wings close to her body, and tossed her into the sky. In that moment of release, birds will always seem to just hang in the air. Then they know they're free. Their wings unfold. I felt that first great wing-beat of the golden as a pure rush of freedom.

Kestrel and I watched. The eagle flew across the valley just above the grasses and up into a tree. We waited fifteen minutes to make sure the bird was OK. We packed everything up, and, as we turned to leave, we saw the eagle drop out of the tree, open its wings, and fly straight at us. It looked down on us, cocked its head once, then flew right over us into the branches of a tall tree next to us.

Kestrel and I had looked up at the eagle. "Thank you, little brother," we said. "Have a good life."

The powwow drums broke through my memories of the release day, and I stepped forward—a man free as he could be of the cancer, a lucky man who knew just how lucky he was.

CHAPTER TEN

SARVEY USUALLY HAS NO more than four coyotes to rehabilitate in a year, but in 2002 we had a boom. Eight of them were rehabbing with us in the coyote pen about twenty feet away from Freedom.

One summer day I was cleaning Freedom's flight. The air seemed almost blue. It was warming up. I knew the planes would be out. There's a little airport down the road a few miles from the wildlife center, with many old World War II planes and new civilian planes, and every time the coyotes heard the planes, they'd go crazy. I heard the faint whine of plane engines overhead. The coyotes let loose with their high-pitched *yip yip yip*s and howls. They were full tilt—about as loud as coyotes get, heads back as if they were howling at the moon, full speed, all out.

Suddenly Freedom joined in. She threw her head back and gave it all she had. I stood there in amazement, almost unsure it was real. I laughed, realizing no one was going to believe me. When the planes left, the coyotes stopped howling, and so did Freedom. It was quiet. I wondered if I was the only man alive that had ever had the chance to witness coyotes and eagles giving a joint concert.

AROUND THIS TIME, FREEDOM'S original anklets, jesses, and leash had begun to wear. I wanted to replace them with something more fitting than the ordinary leather ones we had begun with. I ordered custom-made buffalo-hide anklets and jesses from a catalog, but the leash was another story.

A Diné friend, as the Navaho prefer to be called, had been invited on a buffalo hunt by the Nez Perce tribe in eastern Washington. After the traditional prayer ceremony asking the buffalo's permission, the hunters killed eleven buffalo. The buffalo were distributed among the tribes taking part in the hunt, and every part of the buffalo would be used— the hide, the meat, and the bones. My friend told the Nez Perce the story of Freedom and me. Without a word, one of the hunt leaders gestured to a fallen buffalo. My friend knew what needed to be done.

He crawled into the chest cavity of the dead buffalo and

cut out the heart. He was given the complete hide to take home too. Then a strip was cut from the hide to make a new leash for Freedom.

My friend met me at Sarvey and handed me something wrapped in red cloth. It was Freedom's buffalo heart. I thanked him and the buffalo. Then I took the heart out to Freedom on the red cloth and held it out to her. She stood straight up, her crown rose, and her gazed fixed on it The heart was so big I had to cut it in half. Freedom gorged herself on it.

On his next trip up, my Diné friend brought me the leash he had made. It was well crafted, the leather supple, but strong. He had made it according to my specifications, because I wanted to use it with an old brass clasp that had special significance for me. My dad had given it to me years before, and I had brought it with me to the Northwest. Keeping my dad near us meant a lot to me.

WE BUILT FREEDOM A new flight in front of the wildlife center, moving her from the old one, which was really just for rehabbing, never intended for long-term use, and only had one open side. Kaye gave up part of her view of the river valley so Freedom could have it. The Kwakiutl who live a little farther north on the coast from my home have a story I've always loved about eagles' eyesight. An eagle who'd been posted to watch for enemy canoes traded eyesight with the slug (which at that time had excellent eyesight). The eagle's amazing vision was supposed to be temporary, but after he did his job, he refused to return the vision. I probably would've kept the superpower too. A few trades in my rock-and-roll days went down just like that.

This new flight was on the side of the hill and had a large Douglas fir next to it. Freedom had custom-made perches and her pool. We called it her cathedral.

At first it seemed like she might share it. Yakala was a male bald eagle living by himself. His roommate, going blind and losing her faculties, had been euthanized. Tua's quality of life was miserable without her eyesight, and when she fell over she had trouble getting back up. Yakala kept pooping on her.

A few months after Tua was gone, the director asked me if we could try moving Yakala in with Freedom. We didn't expect the pooping to be a problem with an able-bodied bird like Freedom. I was more worried about what Freedom might do to Yakala. She'd had a roommate once before, when she was not quite a year old. I put the other eagle in her flight and watched both of them for about ten minutes. Nothing happened, so I left. I went back to check on them after five minutes, because eagles are sneaky, and sure enough Freedom had the other eagle pinned on the ground. She was standing on his back as he struggled to get free. I had to shoo Freedom off and bring the other inside to look him over. He wasn't hurt, but we'd never tried it again. Poor kid, he was also less than a year old, and scared.

But maybe as an adult, Freedom could accept Yakala. Neither of these arrangements were in any way related to breeding—Sarvey doesn't breed its raptors and they've never shown much interest. In the wild, eagles basically mate for life, but if a mate dies, the surviving eagle will find another partner. They also seek new partners if breeding attempts fail.

They moved Yakala in with Freedom on a day I wasn't there. We wanted Freedom to get used to Yakala without me to distract her. I checked in by phone for the next four days to see how things were going. The word each time was that they were both getting along. There were no problems. I was surprised, but I guessed I shouldn't have been. Freedom had a way of surprising me.

I drove to the clinic, and when I got there, I went over to Kaye's house. I deliberately didn't walk over to the flight. I wanted to sneak up and get a better feel for the way these two were coexisting. I popped in to Kaye's house and chatted a little. Then we walked across the dirt driveway and onto the grass leading to Freedom and Yakala's flight.

As soon as I got to the netting Freedom turned and saw me. Her crown went straight up and she turned on Yakala. She screeched at him full volume, then jumped down off her perch and started chasing him. Poor Yakala was confused—

one minute things had been kind of normal and the next this crazy bird was chasing him.

Freedom wouldn't stop. She eventually pinned him on his belly next to a large log. Kaye was amazed. She smiled at me. "She's claiming you." I couldn't suppress my smile. Freedom was pure instinct. And I was her human.

We took Yakala out of there immediately and put him back in his old flight in the educational building. He got to see friendly people every day. I would have bet he didn't miss his crazy eagle roomie.

In January 2003, Kaye called me to say that two black bears were on the way to Sarvey from the Washington Department of Fish and Wildlife. The bears would arrive by 10:00 P.M. She didn't know much more than that.

They must be cubs, I thought—we'd never received two full-grown black bears at the same time. After work, I headed for the center. A few volunteers came to help. The bears were due in around 10:00 P.M. so we waited—and waited.

Around 11:00 P.M., headlights appeared. A Fish and Wildlife truck towed a big trailer up to the center. The rig looked pretty spacious for two little cubs. Two WDFW officers got out and started to tell us what had happened. But we wanted to see our new bears.

WDFW opened the back doors of the trailer. You could hear the collective "Oh my God!" from the Sarvey people. The bears were full grown, and enormous. I had never seen a bear as fat as those two. A full-grown black female bear should be about two hundred pounds—heavier in the fall, and lighter in the spring; these bears looked about four hundred pounds.

They were females and had been named Corky and Pumpkin. They didn't move. We had to get closer to see the full extent of their misery. Their fur was dull and brittle; their nails were cracked, some broken off. They had yellow cracked teeth and bright red gums with some bleeding. Worst of all were their eyes. They had such sad, beaten-down eyes, eyes that were pleading for help. They clung desperately to each other.

We talked to them gently, then brought them a couple of bear platters—code name for apples, honeydew melon, filberts, walnuts, almonds, blackberries, blueberries, trout, and more. Their eyes lit up at the smell, and they dove in like they had never had real "bear" food before. Which they probably hadn't—the WDFW officers said the bears had been subsisting on a steady diet of turkeys. No one told the torturers who "owned" these bears that bears don't eat turkeys.

As they were crunching walnuts and eating grapes their whole demeanor changed and relaxed. The cages they'd been living in had no water, just a couple of barely wet buckets.

There were feces and urine puddles on the floor. These were nothing less than atrocities.

And it was nothing less than a miracle that they were still alive; even that showed their sweet spirits. We put in big bowls of water and cleaned around them. They would stay in their cages in the truck until we learned if we could move them to the large mammal pen. We wouldn't want to have put them through the stress of the transfer if they couldn't stay.

We got the story from the WDFW officers. It had begun at the Canadian border a few hours earlier when the trailer Corky and Pumpkin were confined in was pulled over for inspection. The guys driving told U.S. Customs that the bears were going to Hollywood to be in a movie. But a narcotics dog signaled his handler that all was not well. U.S. Customs agents found 166 pounds of marijuana and $180,000 in cash stashed in the false walls of the bear pen and the trailer frame. The value of the pot was estimated at half a million dollars.

Customs had called in the WDFW to deal with the bears, and the Drug Enforcement Agency to deal with the drugs, so there were now three different federal agencies with an interest in these bears. The driver had spilled his guts about his bosses. These punks were using Corky and Pumpkin for cover as they crossed the U.S.-Canada border, smuggling high-grade Canadian pot into California, where they traded for cocaine to smuggle back to Canada. We were to hold Corky and

Pumpkin for five days or until the bust went down, and they had to stay in their cages that whole time, held as evidence in a federal crime.

After the arrests a week later, the WDFW asked if Corky and Pumpkin could stay at Sarvey for the duration of the legal proceedings. That could be up to a year or more. We were relieved the bears weren't being returned to the human garbage who used them for dope smuggling. The large mammal pen was all ready for them. They had three big rooms and a huge tub of water and two wooden houses in a trailer, so they could hide if they wanted.

All we had to do was sedate both of them and move them in a heavy-duty cart into their new home. Since they were obese, we had to keep a special eye on their breathing when they were sedated.

We got them into their new home. Now we had to start earning their trust. Though they were habituated to the scum who had used them, we would need to expose them to a completely different human experience. When we would go in to feed or clean their new home, Corky and Pumpkin would huddle together in a corner, hanging on to each other. It was heartbreaking to see these two massive bears so afraid of what would happen to them.

The volunteers who worked with Corky and Pumpkin took their time. Once the bears figured out we were not going to hurt them, they began to relax.

Corky and Pumpkin loved berries, raisins, and, more than anything, raw eggs. They would use their mouths to gently take the eggs out of your hand. Then they'd sit down, crack their prizes, and slurp up the egg. The first one done would come back for more, with the other not far behind. You had to watch their intake because they would eat as much as you gave them. They would do that with berries too. We were serious about them losing weight. They didn't care.

As spring drifted by, Pumpkin and Corky began to lose weight a little at a time. Their coats vastly improved. Their eyes sparkled once again. And they learned to play.

They pulled off one of the slats on top of one of the houses, and whenever one of them went in we'd see a bear's head popping out to look at us. I called it "Run silent, run deep" for the old Clark Gable submarine movie.

One very hot day in early summer as I was filling their water tub from outside the chain-link fence, Pumpkin started running under the water shooting from the hose. Play time! I brought the hose into the pen. Pumpkin was sprinting end to end and jumping at the water, bucking almost like a horse, so Corky had to get in on this too. I could have sworn both of them were smiling as they ran past me.

Back and forth, the two bears ran jumping at the water coming out of the hose at full blast. In no time, bears, me, their entire enclosure, everything was soaked. After thirty minutes or so Corky took a break, but not Pumpkin. I would just about get the tub filled—which took at least ten minutes—

and Pumpkin would charge over, jump into the air, and land in the tub. There was three hundred pounds of Pumpkin and an inch of water. Finally, Pumpkin decided to sit in the tub and let the water splash right onto her.

She would lift her arms one at a time. I would spray under each arm. She would let the water hit her on her chest and bite at it, then snap her head around. I knew she was having the best time she'd had in her whole life.

I looked over at Corky and she was on top of their house sacking out—because the floor had six inches of water on it. Pumpkin and I spent most of the afternoon playing in the water and running up quite a water bill.

Some of us at Sarvey were now employed as bear caretakers by EG&G (Edgerton, Germeshausen, and Grier, Inc.), a contractor who in turn worked for U.S. Customs. The bill was averaging $3,000 a month just for food and medicine, and EG&G was paying those of us who worked with them too. Corky and Pumpkin were, you might say, in the Witness Protection Program. There was strict control over who could see them. The only ones allowed were the bears' caretakers and five other people from federal and state law enforcement. The officers on the "cleared list" had to sign in and out. Officials from other agencies weren't even allowed to see them without clearance from Kaye Hickman at U.S. Customs. She would wind up being one of the bears' most important and influential friends.

After the smuggling ring was arrested, we got permission to give the media the story. King 5 news in Seattle, one of Sarvey's biggest media supporters, got first crack. The television station ran multiple features on Corky and Pumpkin. Other local news stations did the same. We had TV news plus newspapers from Seattle to Bellingham. Their story ran coast to coast from Boston to L.A.

Then Kaye Baxter asked me to start looking for a permanent home for Corky and Pumpkin. Our bear enclosure was only for rehab. We needed to be prepared before we were asked.

There were two nonnegotiable terms of placement: (1) Corky and Pumpkin must remain together. To separate them would be extremely cruel. (2) They could not be bred or put on display. They had been through too much.

The Wild Animal Orphanage in San Antonio, Texas, took in unwanted, abused, and nonreleasable wild and exotic animals. That organization met our criteria, and personnel there said they could accommodate the girls. Everyone who knew anything about the WAO said the same thing, "Those bears couldn't want for a better place to live."

WAO was the perfect place, but first we had to make sure they wouldn't be returned to their "owner" in Canada who had mistreated them so brutally. Now that Corky and Pumpkin had learned to trust people again, if they were returned to their former abusers, it would be even worse than before, a slow death, emotionally and physically. And the SOBs were trying their best to get them back. I told Kaye I would rather shoot Corky and Pumpkin than return them. If that's all I could offer these two sweet bears, then I would do it. I would have much rather shot their former owners.

Around mid-September we got word I wouldn't have to shoot anybody. Although their torturers wouldn't get what they deserved, they wouldn't get Corky and Pumpkin either. We could find a new home for our friends. I called Carol Asvestes, the director of the Wild Animal Orphanage, and told her the news. She told me it would take $15,000 to get the bears transported to San Antonio and to fund a new enclosure for them. I called Kaye Hickman at U.S. Customs. She promised to call Washington, D.C., and try to get the money from the agency's forfeiture fund. Sarvey held its collective breath.

A week or so later we learned the money had been allocated for Pumpkin and Corky's move. They had a new life coming; everything was in place. It was party time at Sarvey. It almost seemed unreal.

It was late October. We had a small window of time to get the bears rolling. The WAO people were going to New Jersey in December to rescue twenty-eight tigers, and they needed Corky and Pumpkin settled in first. It would take at least a week or so for the humane train to get to us and then tow a huge trailer with two bears from Seattle to San Antonio, and of course we had to worry about winter weather slowing the trip.

Getting them ready to leave started the night before.

First we had to sedate them . . . and Corky was first up. After Pumpkin saw this, she knew what was coming and tried to fight getting stuck with a pole. Her adrenaline was pumping, and it took longer than we originally thought. Then we had to move three hundred pounds of dead bear weight, twice, into their traveling pens. The whole thing took several hours. It was exhausting.

Once safely in the transport enclosures they would stay the night. That way we could check on them throughout the evening to make sure they were all right.

All the major TV news stations from Seattle came the next morning to see them off, along with radio and newspaper people. Kaye from U.S. Customs, the head of their publicity department, and a few other federal and local law enforcement officials, plus many of the volunteers who helped take care of Corky and Pumpkin, were all there. It was a zoo.

Pumpkin and Corky were still a little groggy, but also nervous about being uprooted once again. We knew this was the last time they would have to do this, and this was for the right reason. We told them it was going to be OK. One of our most dedicated volunteers, Sandy, made them her famous "bear bars," baked with sugar and eggs and other stuff, a road treat she had created especially for Corky and Pumpkin's long trip. We loaded the bears up with raw eggs too. A bear can't travel without eggs.

The time came for them to leave. We told Corky and

Pumpkin not to be afraid, just to trust us; they were going to a great new home. We gave them a last pat on the paw and a last nose to nose between the links of the traveling pens. It was a relief to see them go where we knew they'd be safe, but losing two dear friends was wrenching. Many of us cried.

It was another relief when we heard the journey had gone OK. Carol at the Wild Animal Orphanage said they'd stopped in Colorado when it was snowing and opened the side doors so the two bears could see and smell the snow, but it was otherwise uneventful. On arrival they had thirty days of quarantine— always necessary before you release newcomers into a closed population—but after that, they joined several of WAO's black bears, a few sun bears, some grizzlies, and other bears.

THAT DECEMBER, OUR RESIDENT COUGAR, Sasha, went off her feed. Then two days later she completely stopped eating and couldn't stand up.

We knew Sasha was dying, so we didn't need the help of a veterinarian. All we could do was make her comfortable.

It was a wickedly cold and damp December day. Another volunteer and I brought Sasha inside. We had laid her on some warm blankets since her body temp was rapidly falling.

Cats will give you a nibble in friendship. Sasha was lying

with her head on my thighs facing me and she gave me one of those love bites. She was saying, "Good-bye, dear friend."

Two hours later, Sasha died in my arms. We were doing CPR on her. Sue, the clinic director, was breathing for her; Kaye was doing chest compressions, and I was draining Sasha's lungs by lifting her up. But I felt her life go out of her body. She was gone.

I held her still warm body and remembered how she had once given me a little harder love bite on my chest when we were playing. I was bruised for a week with her tooth marks. I wished the marks had stayed.

IT WAS A SAD TIME, but keeping up with Corky and Pumpkin's progress through Carol at the Wild Animal Orphanage was cheering. The bear buddies were having the time of their lives in Texas. In the summer of 2004, Kaye Baxter and I went to the WAO to see our old friends and WAO's tigers, chimps, and other wild ones. The first day it was raining and hot and muggy, which is no joke in Texas, and we couldn't go see Corky and Pumpkin, but the next day was fine. Mary and Michelle, who work and live at the outdoor site, were there to meet us. They had a little golf-cart-on-steroids to drive around the trails, so with the mud flying and the sun shining on a sweltering Texas day, we went off to see Corky and Pumpkin.

As we pulled up we could see two separate big enclosures for about thirteen black bears.

Corky and Pumpkin now had trees to climb, room to run, a big swimming pool, and great people to take care of them for life—and they had each other. Carol had told me that when Corky and Pumpkin got in the pool they wouldn't let any others bears in.

When we got there, I started looking for Corky's telltale white spot. I didn't see either bear.

Finally Carol said, "Jeff, Corky's right in front of you."

It was Corky. She had left her group of bears and walked right up to me. Her fur had turned orange red with a black undercoat coming through. She was absolutely beautiful. All the excess weight was gone, and her eyes shone. She looked like a healthy and happy bear should.

Pumpkin was having a small spat with another bear. They were yapping away at each other. Pumpkin was black as night, trim, and every bit the character. She looked amazing. I would never have imagined that the fat, miserable, tortured bears of eighteen months ago would look like she and Corky did now.

Corky recognized me right away. She looked me in the eyes and sat down no more than three feet in front of me. Then she raised her right paw and extended it toward me as if to say, "Hey, I've missed you. And thanks."

I was speechless—and grateful. Corky still had her paw

extended as we just looked at each other. I was with my old bear friends. At that moment everything was good in the world.

I called on that memory to comfort me later that summer, when Lynda and I lost someone who had been dear to us for a long time. Eight years was not long enough with Mr. Timms. But, we didn't have control over that. By late summer, Mr. Timms was not stealing as much food as he had. In fact, his appetite seemed to be fading away. He grew thinner and thinner. He no longer ran on the squirrel highway or much of anywhere else. In his last days, he let Lynda hold him again like when he was a youngster. It was bittersweet for both of us.

Finally all he would eat were tiny bits of watermelon from my fingers. On his last day with us, I was feeding him and he peed all along my arm. He'd never done that before. I knew he'd lost control of his bladder. He was near the end. Lynda and I knew we had to grant him our last act of love.

I took him to our vet. There, I held Mr. Timms and put the gas mask over his face. He struggled for a second, but I knew what had to be done so I kept gently holding him. Once he was unconscious, the vet injected the drug into his heart and he was gone peacefully.

We had him cremated and kept his ashes in a little wooden box in our living room. And the day I took down his house was one of the hardest days of my life.

CHAPTER ELEVEN

I T WAS THE HEIGHT of spring 2005. But it wouldn't have mattered if the weather had been a blazing 120 degrees, raining boulders, or sleeting icebergs, because that day I was told the best news a former cancer patient can hear.

I drove to Virginia Mason to see Dr. Jacobs. As I walked in the front door, I wondered if I'd come to the wrong place. The first thing I saw was a water wall. Someone had to tell me where to go. Check-in, the blood lab, and the infusion clinic were now on the first floor. Everything was shiny and new.

Call me crazy, I missed the old place. So much had changed. I had grown used to the shabby twelfth floor. It had a lived-in quality. But I guessed I was grateful I hadn't been in treatment long enough to watch the change. I headed to Dr. J.'s office.

He welcomed me with an ear-to-ear grin. "Congratulations. You don't have to come back here anymore. Just get a physical once a year."

I thanked him. He wished me luck. There wasn't much more to say. We shook hands and I was out the door. I was headed for my truck when the Christmas card I had sent him in 2000 came to mind. It was a picture of Freedom and me. And I had written to him: "Thanks for saving my life."

My life after the good news mostly seemed ordinary. I went to work. I hung out with Lynda. She gardened. I took photographs. Saturday nights, we went to our favorite Mexican restaurant. Weekends I went to Sarvey and took Freedom out on the glove for walks. Maybe talking with an eagle wasn't that ordinary—but it was to me.

I missed my friend and co-conspirator, Crazy Bob, who had died a month earlier. He had told me in 2004 that he was going to have to stop our ambulance runs. His chronic illness had worsened to the point that too often, he couldn't muster his strength. Still, he had made a few valiant efforts in his last year—especially if it was for a raccoon. In the times I visited him, we had often reminisced about one of our favorite runs.

I had gotten a call at home from the guy who took care of the greenbelts near Lynda's and my house. He'd seen a hurt raccoon and he thought it might be in a trap. I walked deep into the greenbelt checking out every fallen log and poking into tall bushes. I didn't see the raccoon. I made two trips through to be sure, then gave up and walked back home. Just as I stepped out of the greenbelt, I turned around for one last look and a gray and black head popped up.

I walked slowly over to see what was going on with the raccoon. He was terrified and furious. I could see that his two rear legs were in two separate leg-hold traps. There was blood everywhere. Leg-hold traps are illegal in Washington State—for good reason. They are absolutely vicious devices. The worst design snaps down on a wild one's leg(s) with sharp metal teeth. Wild ones will attempt to gnaw their own paws or limbs off to escape—sometimes they succeed.

This trap didn't have the teeth, but it had tortured the raccoon to the limit of his endurance. Both legs were broken. I looked at him and wanted to inflict the same damage on the monster who had set the trap. But there wasn't time to dwell on that. I knew it was a bigger job than I could handle. It would take two people, and I knew who I wanted as the other person.

I backed away from the raccoon slowly so I wouldn't frighten him more and rushed home to call Crazy Bob and ask if he could come out of retirement. "I'll be there," he said.

I called the cops, and then I called the Sarvey ambulance to ask for transport once we'd gotten the raccoon free.

Bob was at my house in fifteen minutes. Three cops in three cop cars showed up right after him. We all walked back to the raccoon. Bob and I scoped out the situation. There was a big downed log right next to the raccoon and big bushes right around him. We got our plan together, got every detail in place, and then we went and got our technical equipment—which consisted of a net and a carrier.

The wily cops stayed a safe distance from the raccoon. They were in full raccoon alert, eyes bugging out, their hands on their guns, their faces set into stony masks, their entire aspect in cop overkill mode.

Bob used a variation of his unique and foolproof "scoop and pin" technique. He pinned the raccoon under the net. I opened the jaws of the traps and freed the raccoon's legs. In a millisecond Bob had flipped the net over, under, over again, and pinned the netted raccoon with the metal net frame. And I had tossed the traps back to one of the cops.

Since Bob was on the ground with a netted raccoon on his hands, he couldn't get up. I grabbed him by the back of his pants and wedged him upright. Bob put the raccoon in the carrier. I slammed the door shut. It had taken thirty seconds. I looked up. The cops were frozen in place. A big crowd had gathered. One lady glanced at the cops and gave Bob and me a thumbs-up.

To Bob's great joy, a friend who worked in the police department told him later that the cops had come back all fired up. "We just saw two guys free a trapped, pissed-off raccoon. Those guys are crazy. They shouldn't be out on the streets."

When the raccoon arrived at Sarvey, he was taken to the

vet. The doc amputated one of his legs. The surgery was successful, but the raccoon was fighting a massive infection. He went into a coma for a week and then recovered. Except for his temper. But he was released in a safe place; raccoons are OK with three legs.

That was one of Bob's last calls. He finally grew too weak. And then, on February 11, 2005, Bob Jones, aka Crazy Bob, lost his battle for his life. He was sixty-one years old. I'd gone to be with him as soon as I received the call from the hospice nurse. I sat with him for a while, then went out into the living room and waited, thinking of Bob's many escapades. He was too young to die. Then the nurse came out of Bob's room and said, "Please come back now. It's time."

I went in just as Bob was leaving. I looked at him and thought about who he'd been, his legacy to the wild ones, and how much truly crazy fun and heartbreak we'd had on our ambulance runs and rescues. I looked back on what he had given—all of himself. He had constantly put himself in harm's way and showed up at Sarvey anytime day or night, bloodied from the battle but always with the patient in tow. Even if that animal had to be put down, Bob always thought it was better than letting the little wild soul die alone, cold, and with no one to tell them, "I love you."

I took one last look at him. I knew he was hanging with his raccoon buddies on the other side. I hoped he knew how grateful I was to have known him.

Next time I saw Kaye she told me for the first time just how significant Bob had been to Sarvey. I'd known for years that he was very wealthy, but not that there had been times years ago when she'd almost had to close Sarvey the next day due to lack of funds. On those occasions, Crazy Bob had given Kaye thousands of dollars under the condition that no one would know about it.

"You know," Kaye said, "we just lost one of the best friends Sarvey and the wild ones will ever know."

ONE OF THE FIRST Sarvey residents that would never know Bob came on July 25, 2005. The Sarvey ambulance picked up a young eagle that had fallen from its nest in Olympia. The bird was huge, big even for a female. She had multiple fractures of the humerus bone in her right wing. The fall had compressed the fractures so the wing could not be pinned. We had to wrap it and keep it immobilized. The wing healed well enough so that she could fly, but not well enough to be released.

While she was inside healing up—before she could go into a flight—she began to engage in an unusual behavior. She liked to play with anything—a towel, a bowl, even her food, whatever was lying around. One day a volunteer brought in a squeaky ball and rolled it to her. The eagle pounced on it. It squeaked. Her eyes lit up and she cocked her head. "What's this?" her face said.

She grabbed it with her talon. One squeak was all it took. She started jumping up and down, making the ball *squeak, squeak, squeak* as she hopped all over the back room. She'd even spread her wings getting lift for the jumps. Those of us watching were doubled over laughing. There was that giant bird jumping up and down with a ball in its talon making squeaking sounds. She'd let it go. One of us would grab it and roll it away from her. She'd set off racing after the ball. Pounce. *Squeak.* Over and over.

Shortly after that, we named her Wanbli Askata—eagle that plays.

We moved her into the flight next to Freedom's.

Wanbli Askata never vocalized until she started living next to Freedom. After she learned how to talk, she wouldn't shut up for the longest time. Freedom had a hard time getting a *chak* in edgewise.

Freedom still came to me in dreams. One morning, I woke and said to Lynda, "I had a dream that Freedom was standing on my head."

She laughed. "You and that bird!"

Later that morning, I got a call at my job: "Freedom has hurt her wing and lost a lot of blood. She's scared and needs you here." I raced out of work, shouting back over my shoulder, "I gotta go—the kid needs me." Everyone at work knows who my "kid" is.

Talk about your heart in your throat. I jumped in the truck and laid rubber. Kaye had received the same call and she was just as far away in the other direction. There we were, both hauling ass, doing well over eighty miles per hour to reach Freedom. I got to Sarvey first in a cloud of dust and screeching tires.

I climbed out and ran into the hospital. Freedom was in a large green cage with her hurt wing bandaged. The instant she saw me, her posture relaxed. I opened the door and held out my gloved arm. She stepped gingerly onto my arm. We stood there, beak to beak. "How are you?" I asked softly. She let me know she was fine.

I stayed with her as we walked around the clinic for a while. I could feel both of us settle down. Sue, the clinic director, told me what had happened. A tiny bump on Freedom's perch had worn down to a point, and Freedom had hit it just right with her crippled wing. The sharp piece of wood had hit a major vessel. A volunteer had spotted all the blood and called for help. They'd treated her right away.

A couple of days after the accident, I was at Sarvey when Sue checked up on Freedom's hurt wing. Freedom was healing well. Then it was my time to "rescue" her. I always "rescued" Freedom from the big bad people who were doing mean things to her at the end of an exam or coping (trimming talons or beak, which grow like fingernails without the natural wear of living wild). I would get her on the glove and ask her, "What have these people done to you?" and she would proceed to tell me all about it in great detail.

As I finished my part of our routine and before I could put her jesses on, Freedom stepped up to my shoulder, then to the top of my head.

I couldn't believe it—the dream had come true. I was in shock. The rest of the staff was thunderstruck; it was such unusual behavior—and they didn't even know about the dream. Like the time she embraced me, it seemed to be a unique event. I reached up with my right arm, Freedom stepped onto it, and I brought her back down. Then she gave me a look that I knew meant "I know what you dreamed—because it was me." It was an ordinary miracle, not like saving my life, but no miracle is really ordinary. It reminded me that Freedom's extraordinary gifts are more than I can really understand—though they always have my awe and love.

Even though the wound was healing nicely, Freedom had to spend a few weeks inside so it would heal completely. We examined it every other day, redoing the bandage as necessary and taping her back up.

Freedom was a model patient—except when it was time to clean her dowel cage. Volunteers would conveniently have something else to do when that chore came around. She could be very intimidating, especially to newer volunteers. We'd let her out first—but even that was an opportunity for trouble. Freedom had the ability to sense fear and uncertainty and act on it. In her flight she would go after people who showed fear around her, chasing them as they ran away. Or she would pin a volunteer in the corner of her flight with her wings out-stretched—barring him or her from leaving.

One time while she was still living in the dowel cage after the injury, I'd taken her out for her walk, brought her back in, and set her on her perch in the cage. I gave her a chunk of salmon, which she promptly threw on the cage floor and ignored.

Another longtime volunteer, Tommy, who had been gone for a couple of years, came by. He was back in town and wanted to start working at Sarvey again. Since he had heard about Freedom, he asked me if he could meet her. She took one look at him and tried to attack him through the dowels. Her talons slammed between the bars and almost nailed the guy. She flapped her wings violently and screamed at him. All her feathers were puffed out, and she had a wicked gleam in her eye. He jumped back so fast he nearly lost his balance.

I was as surprised as Tommy. I had never seen this side of her before. And I was a little proud too, of how instinctual she was. *That's my girl*, I thought. She never did this with any other volunteer, and I really don't know why she did it with this guy.

Fortunately, Freedom could move back to her outdoor flight after two more weeks. But there was another incident soon after she was back in her flight. I was out in front of the clinic one morning when I heard a faint "Help me—help me!" coming from somewhere out back. I followed the sound to Freedom's door. A young volunteer was cowering in the back corner with Freedom lurching back and forth in front of her, wings spread wide, head up, giving her what for.

I tried not to laugh, but I knew Freedom was just a big bully, lots of blather and no bite. "Just walk around her," I said. "She won't hurt you." The volunteer tried to do that, but Freedom jumped in front of her and menaced her back into the corner.

I had to walk over to Freedom and say, "Please let this young woman go. She has other flights to clean and you're holding her up." Freedom looked at me, turned her back, and walked away. I got in between them.

The volunteer bolted out of there, never to return to the flight. She never went near Freedom again. I knew that if she had been able to let Freedom know that she wasn't afraid, Freedom would have picked up on it and would have wanted to play. That was really what the chasing people was all about. She wanted them to chase her too. She loved party time with people.

NEWT, A BARN OWL, did not look like a star. But I could remember a time when Freedom didn't look like one either. He was brought to Sarvey in a cracked egg in 2005. His egg had fallen from the nest in a barn—of all places.

After Newt hatched, he looked like a tiny fuzzy dinosaur—a tiny fuzzy pink dinosaur. He was about two inches from head to foot. His head was huge, at least three-quarters of an inch long. The minuscule feathers on his head looked like a double Mohawk. If you'd found him at the base of a tree, you couldn't have guessed what he would become.

A few weeks later, we set up Newt and a couple of other baby barn owls in an incubator that faced the back of the dowel cage. We put Barney, an adult barn owl, in the dowel cage long enough for the baby barn owls to watch him. Baby wild ones need to see an adult of their species from the instant their eyes open and focus. It's called "imprinting" and is critical if they are to know what they are.

After two months of a steady diet of minced mouse, then whole mice, Newt's face was the distinct pale disc of a barn owl. The rest of his body looked like a big cotton ball. He was about eight inches tall, his eyes were shiny black, and his legs and feet were pale cream.

Newt lived at Sarvey with his young owl people for a few months, then all of them were released. We had put them in the hunting flight one by one with live prey to make sure they could survive on their own. Each owlet passed its final exam, and they were taken to their new homes—farmland, forest meadow, river tree line, anywhere they would have prey, a roost, and space—and set free.

Transformation of an infant wild one into an adult is always miraculous, but with Newt we got to see a tiny fuzzy pink dinosaur become a perfectly engineered predator. His facial disc was fully formed, a pale halo around his eyes and beak. Barn owls' facial discs are not for show. They funnel sound into their fine-tuned ears. Their flight is the most silent of all the owls. As they close in on their prey, they shut their eyes during the last few feet, and their acute hearing guides them in for the kill.

It is believed that many ghost stories originate from the barn owl in full flight. Many of the First Nations believe the owl is a messenger of death. Newt, on the other hand, was a messenger of transformation—and a reminder to me of how transformative Sarvey can be.

About that same time, a Canada goose came to us. Fishing line was wrapped around its leg and had cut so deep that the flesh was growing around the line, and the leg was getting infected.

Kaye and I and another volunteer took the goose into the medical room. We had to cut its skin to remove the line. The line was embedded so deep that in pulling it out, we cut a major blood vessel. The goose was bleeding all over the place. His heart stopped and he quit breathing.

Kaye wrapped his leg in a hunk of gauze and told a volunteer to hold it tight. She began CPR, chest compressions first, then alternating compressions with mouth-to-beak breaths. I pressed the stethoscope to his heart. In a minute or two, I heard the thump of his heart begin again.

Kaye sutured the vein and wrapped his leg. We watched

him come out of the anesthesia. Then he was put in a large green and watched carefully over the next couple of weeks. After the leg was completely healed, the goose was released on his home turf. I watched his big wings carry him away toward his new life, and I was grateful to have been part of bringing him back from death.

We also had a spotted owl as a patient for a few months. He loved to sit outside his large green and we let him. This owl could not be released because of damage to a couple of his talons. He could not close part of his right foot, therefore could not hunt and would starve to death if released. He loved to hang out peacefully in the window of the big cage room.

On one cold, wet, and dreary Thanksgiving I was at the center with a handful of volunteers. We'd arrived early to feed and clean so that people could get home to be with their families for Thanksgiving dinner. I stayed late to help with a bald eagle that had been shot.

This girl had seventeen shotgun pellets scattered across her wings and chest. If that wasn't bad enough, she also had a severely infected foot as a result of the shooting. This needed to be drained, repacked, and rebandaged.

I was to be the muscle and hold her down while Kaye worked on her foot. I caught the eagle and took her into the medical room. Five of us were in there—including the eagle—as Kaye began to work on her foot. The center was quiet.

We were moving right along working on the eagle when

a strange and incredibly beautiful sound came softly from the hallway. We froze and looked at one another. I had someone hold the eagle's talons for me. I walked slowly to the corner and peered around.

I saw the spotted owl on the window ledge looking outside. He was singing his heart out. I went back to the med room and told everyone exactly who was doing the singing. It was our brother, Spotted Owl, and he was singing his song of life. We all smiled at one another. No words were necessary.

His sweet voice filled the room for a long time while we continued working. There we were, Thanksgiving Day, trying to help this damaged eagle, while our owl brother sang us his spirit song. That was one of the best Thanksgivings I ever had.

CHAPTER TWELVE

I N SUMMER 2006, KAYE said to me, "I have to go to Virginia Mason." We both knew what that meant. She had been ill for quite a while, and it was getting hard for her to breathe. All I said was the echo of her words to me six years earlier.

"Damn."

I thought back to my own cancer and Bob's. Sometimes I had this superstitious feeling that it wasn't a coincidence that Bob, Kaye, and I all had cancer and were all drawn to the wildlife center. But I knew it really was just a coincidence. Many Americans get cancer every year, and we are just part of the stats.

Kaye had terminal lung cancer. She had two years to live maximum. I was about to lose a good friend and a great mentor.

After Kaye's dire news began to sink in, my next thought was for Freedom. I had to save her. If we couldn't secure funding to buy the land that Sarvey sits on after Kaye's death, it would be sold to the bank. Sarvey would be gone. I couldn't let Freedom go to another rehab center. I couldn't bear the thought of being separated from her. That would be crushing for me—and for her too. I knew that when eagles are separated from the person they have bonded with, they can go into deep depressions.

All of us were there because of what Kaye had done. I had a special relationship with a bald eagle because Kaye had started that place. Thousand upon thousands of wild lives had been saved because of Kaye. If she was no longer there, what would happen to it—and all the wild ones that come through here, the ones that are saved and the ones that get a merciful death to end their agony? What would happen to the damaged people who find their self-esteem and a purpose at Sarvey? I had many questions and no answers.

When Freedom was a very young bird I had made her a simple promise. I told her "I would do anything for you." Now I would deliver on that promise.

Kaye owed money. She had used her life savings to keep the place going. We had a phenomenal group of people running the center, well-trained people, people who knew the complications of working with wild ones, but we didn't have the contacts in the world of big money.

Every time I talked with the staff people, the only topic of conversation besides our patients was what could we do to save Sarvey. None of us had an answer, though we knew we couldn't let the place go under. The impact of that would affect thousands of wild ones. We all knew stories of worthwhile organizations that had fallen apart when the founder died. We knew we needed to have a different future.

Crazy Bob saved Sarvey again. Although hardly anyone at Sarvey knew his secret—he was extremely wealthy—he had told me years ago; and in 2001, he had asked Lynda and me and retired vet Dr. Myron Phillips to sit on the board of directors of a charitable trust he was founding before his death. Its job was to dole out a percentage of the profits made from his investments each year to nonprofit wildlife rehab centers.

I asked Myron and Lynda what they thought—could we buy the land to keep Sarvey running? The answer was an enthusiastic yes. The trust donated the money for Sarvey to buy the clinic's five acres after Kaye's death.

Kaye continued to be an active part of the center until a year before her death. So much was hard for her. She had lost Mellow Yellow, her beloved red-tailed hawk, to death a few years earlier. And, in that last year of her life, she would sit by her big window in the Hawk's Eyrie, as she called her home, and watch all the comings and goings at Sarvey. She told me it was heartbreaking to see us headed out for programs with our birds and know she no longer had the strength to handle a raptor.

Sometimes I'd be taking Freedom for a walk and stop by Kaye's. She'd look out the window. If she didn't, I'd hold Freedom way up high. She'd flap her wings and Kaye would see us. Her face would light up instantly. She'd haul herself out of her chair and make her slow way out to be with us. Kaye would lean in toward Freedom so Freedom could tug on her hair. Will, a volunteer and eagle handler, also brought Freedom's friend, Wanbli Askata, for visits.

Those eagle visits were the only times I could glimpse the old Kaye.

By June 2008, Kaye was fading fast. The very last time she sat with me outside, she told me she could feel the cancer in her brain. It was causing her more problems every day — names wouldn't come, she couldn't remember what she was doing, sometimes she'd just bust out crying. She knew the end was near.

On July 4, 2008, I headed out to pick up Freedom from Sarvey. People in the area set off fireworks. The fireworks terrified Freedom to the point that if we left her alone, she could injure herself running into the walls of her flight. Bright lights raining from the sky are not an eagle's friend.

So on the Fourth of July, Freedom always came home with me. She was an excellent houseguest. I'd put her perch in our living room and she'd sit there looking unbelievably huge—

an eagle in your living room seems about five times the size of an eagle outside. When the sun goes down, so does she. I put her carrier at the foot of our bed and she sleeps all night, doesn't make a sound. You'd never know she was there. She's also very patient. She waits for me to take her out back to get out of the carrier and stretch out, which allows me to clean out her carrier from the night before. It's cool having her at home with us, though I have to keep her comfort in mind at all times and stay with her when she's outside.

I drove to Sarvey that Fourth on what we called the highway of death, a two-lane that is often a fatal road for wild ones and humans alike. I saw a deer grazing on the side and said a special prayer for him to be careful. "This is not a place for you, little brother."

Then I was past him. The drive up was one of those warm summer nights with cumulus clouds hanging over both the western and eastern mountains. The sky was a lazy blue.

I thought of Kaye saying how much she loved living here because we were protected and hidden between two mountain ranges. I would always answer, "Whenever I'm away from here and out in the wide open, I feel like a rabbit with no hole."

I had the truck windows open. The sweet evening air filled my senses. I was not thinking about anything, just enjoying the gift of summer.

As I pulled into Sarvey, I saw that two of Kaye's sons were there. I figured they were just visiting.

I went to see Freedom. She was running around screaming and flapping her wings. She couldn't seem to settle down.
None of the other birds was behaving that way. And the fireworks hadn't started.

Kaye's son came out to me. He said, "If you want to say
your good-byes, now is the time." I wasn't ready for that, but
I knew then why Freedom was so upset. Kaye had been with
Freedom from the beginning. She watched Freedom and me
learn to work together. She watched me help save Freedom's
life—and she watched Freedom save mine.

I knew Freedom knew something bad was happening and
she couldn't help. I needed to get Freedom away from Kaye's
death, but first I needed to see Kaye. I jessed Freedom up, put
her in her carrier, and went into Kaye's house.

She was lying back in her chair at the window. The cancer had robbed Kaye's brain completely. Drugs were keeping
her sedated—without them she would thrash around. She
was in Cheyne-Stokes breathing. It got heavier. I knew the
end was close.

I sat down next to her. I held her hand. "I love you," I
said. "Now it's time for you to move on. It's OK to go. We'll
take care of Sarvey." She did not respond. I waited for a little
longer, but I knew I had to get Freedom out of there. And I
knew that Kaye would understand.

I left the house and went to my truck. The sky was growing dark. Above and beyond my sadness for Kaye, I felt uneasy. I drove carefully on the winding roads to the highway

of death. The sky was full-on dark. The highway is one lane each way, and there is no divider. The speed limit is fifty-five, but everybody does seventy.

We cruised along. Memories of Kaye filled my mind. Cars and trucks sailed past me one after another, sometimes at what seemed the speed of light. We rounded a curve and were on the straightaway headed home. The darkness was calming me. I was starting to come to grips with reality. I was able to think, *This day is the day Kaye dies.*

Suddenly I saw a car coming, closer, closer. There was an explosion of fur. The body of the deer that I had warned a few hours earlier flew into the air and corkscrewed end over end. I saw the moment in slow motion. It seemed to go on forever. The deer's head swiveled as if it wasn't really attached to the lifeless body.

The car headlights illuminated a halo of fur around the deer and the car. I screamed at no one and everyone. I found myself screaming at my deer brother, "Why why why!!! Goddamn it, I told you to be careful!!"

I was furious at the deer for not listening. But then it hit me. Kaye had just died. I knew it beyond a doubt. She was gone. It made perfect sense in a strange and brutal way. I don't know how I knew this, but I did. And it made me even angrier.

I had Freedom with me, so I concentrated on getting her safely to my house. The phone rang the second I walked in the door. It was Leslie, who had become Sarvey's director after Sue McGowan. "Kaye's dead," she said. She told me when she'd died. I looked at the clock and saw that it was at the exact moment the deer had died.

IT FELT STRANGE GOING to the Stillaguamish River Festival a month later. Kaye would not be with us for the first time in ten years. I felt sad, the way you feel when change has washed some important landmark downstream. The festival and pow-wow are held in a gorgeous little valley. Coming down into it, there are more greens than you can name, and on this drive, the weather was perfect—about seventy degrees, gentle sun, a little breeze coming in through the truck windows.

Freedom and I were ready to go. She was in her carrier. Her iron perch was packed. I could feel myself settling into ceremonial time. The drive over went by quickly. By the time I got to the festival, some of the other Sarvey folks were already in our tent setting up. Kestrel had the golden eagle, Hu Iyake; Will had the other bald eagle, Wanbli Askata. We got our birds out of their carriers, jessed them up, and put them on their perches. We set up tables with center information and waited to be escorted into the powwow circle.

The crowd was forming, the dancers mingling with the others. The colors of the dancers' regalia were brilliant—

scarlet and turquoise, yellow and orange. They carried eagle feather fans and wore eagle feather bustles. Bells on their leggings rang silvery in the sweet air. I walked down to watch the human display moving in front of us. Besides, there was salmon, fry bread, and Indian tacos.

At the powwow, the Stillaguamish welcome many other Original Nations. There were dancers from many places and peoples. There were the dancers in jingle costumes, soft leather dresses hung head to toe with little tin cone-shaped bells. There were the grass dancers, whose movements mimicked the steps of dancers of a hundred years ago who flattened the grasses before a ceremonial dance. The hoop dancers weave their incredible patterns with such flawless grace that you think they must be magicians.

Then it was time to begin. The Sarvey bird handlers took our places, me with Freedom, Will with Wanbli Askata, the other bald eagle; Kestrel with Hu Iyake, a golden eagle; and Sharon with Wi Waste, a red-tailed hawk. Our escort, in full regalia, came to take us to the formal powwow circle. They protected the birds on all sides, one dancer in front, two on the sides, a fourth watching our backs. They eased our passage through the eager crowd. I put Freedom on my arm and moved forward. I looked up at the sky and saw a double rainbow in the west.

My senses were flooded. I could smell Freedom's familiar scent—a dry perfume, sweet and airy. There was the pungent smoke of the salmon barbecue. There were brief showers of light rain, so I was breathing the green perfume of the grasses and trees. And, almost enough to make a grown man carrying an eagle break rank, the seductive scent of fry bread cooking.

We had permission to walk across the powwow circle, though the proper way to move into the circle is by going around and coming in. Dancers and military veterans moved in behind us. The eagle staff leader took his place in front of us. The drums began.

We danced in place, a slow step step step step. I felt the drums in my head and heart more than I heard them. The bells and deer hooves on the dancers' regalia kept pace.

The eagle staff carrier began to step forward. We had to wait till we got the word to go. I had a flash of Kaye a year earlier. She had been in a wheelchair, in her full regalia with a bandana tied around her chemo-bald head. She'd carried a red-tailed hawk. I remembered the intensity of her expression.

A cool breeze washed over all of us. The drums seemed to intensify. We stepped around the circle. No rushing here. The grass dancers spun and stomped. The women jingle dancers sidestepped in circles as they moved ahead, the tiny cone bells on their dresses ringing softly. Some of the old men danced simple steps, their feet moving on the earth as though they had danced that way for centuries.

We circled back to where we had begun, making a sharp right turn, and stepped across the powwow circle to maybe twenty feet in front of the drummers. We stopped. The leaves in the trees behind the drummers flipped in the breeze, first their dark sides, then their light. From the corner of my eyes, I could see the dancers and military veterans in a ribbon of motion and color.

The wind began to intensify. I could smell dust swirling up from the dancers' feet. Though it was just past noon, dark began to ascend from the horizon. The trees were shaking, the leaves shuddering bright/dark, bright/dark.

I felt Freedom open her wings and knew her whole body was moving in a way I had not felt before. It was as though she had stretched her wings out and then begun to curve them back in. Both talons ratcheted down on my arm. For a few seconds, I didn't look directly at her. I saw from the corner of my eye that her neck seemed to stretch, her head grow bigger, then her body—bigger and bigger and bigger.

Then I looked at her. She was huge. I saw in her eyes a transformation, as though she was shifting not into something else, but into something both greater and more herself. I knew she was aware of whatever was moving through her. And that she welcomed its presence.

Jolts of raw nerve energy ran down my spine. Every hair on my body stood on end. I felt as though I was plugged into a 220-volt outlet. There was no past. There was no future. There was only "Now."

The drumming stopped as the song came to its end. The wind died down. The sun came out. A holy woman from a

First Nation took the microphone and began the opening prayer in her language.

Kestrel, Will, Sharon, the birds, and I were escorted back to the Sarvey tent. I could hear the powwow starting behind us. I set Freedom on her perch. Kestrel set the golden on her perch. Kestrel's eyes were huge.

"Did you see anything?" I said.

Kestrel nodded. "I was afraid to say anything."

Sharon stepped closer to me. "Did you see that?" I said.

Sharon nodded. "That was Kaye saying good-bye."

CHAPTER THIRTEEN

I HAVE NEVER MET MY friend Gayle Hoenig in person. She runs a small wildlife rescue place in Colorado with her own money. Mutual friends in the world of wildlife rehab introduced us on e-mail sometime in 2004, and we phone occasionally, talking shop and wild ones. Gayle's the kind of person who uses e-mail to full-tilt—I get action alerts about animals in trouble, whether it's dogs, or elephants, or seals in South Africa. She has a huge heart and is a great ally in the work.

I had sent her a couple of photos of Freedom and me over the years. In March 2008, she asked if she could send them out to her e-mail list. "Sure," I said, "but there's a story that goes with them. Would you like to read it?"

"You bet," she said. "I'd love to."

So I took five minutes at work and hammered out an e-mail. I sent it off to Gayle and promptly forgot about it. When Gayle said she wanted to forward my e-mail to some friends, I really thought that would be the end of it.

A few days later, Gayle forwarded me a few of the responses she had received. Within a week I had e-mails coming in directly from utter strangers to both addresses that were included on the message. (I had forwarded Gayle's request from one of my accounts to the other and not bothered to tidy it up, so both of my personal e-mail addresses were included.)

Within a week or two I had hundreds of messages. It didn't occur to me to count them. They started coming at a faster pace, first twenty a day, then fifty, sixty, eighty, until finally I was receiving over a hundred e-mails per day. Finally all my e-mail boxes were full and rejecting all new e-mail. I opened another account with more capacity and forwarded both addresses there.

I could see that something spectacular was happening. I began to see all these strangers almost as a movement, even though they were acting individually.

People wrote to tell me about their own experiences with cancer. "I just was given the magic words CANCER FREE last week," wrote a woman, and I knew she was a sister across the miles. "What I wouldn't give to have a friend and companion like Freedom," wrote a man with inoperable lung cancer. "My friend sent me your story and [I] just wanted to tell you it made me smile," wrote a woman who, at forty-one, had been told she had five years to live. Another woman wrote from England, "When I read this miraculous story I emailed it to my husband who has cancer and is [in the hospital]. The staff at the Oncology section of the Brisbane Hospital . . . photostated the email and have it on the wall for everybody to read." I could imagine that place, a British twelfth floor, and I remembered how much little bits of hope meant to me on my own cancer journey.

People wrote about their own stories with their pets. One woman wrote about the cat she rescued from a concrete quarry when it was a kitten. She cut the concrete from its paws, washed the fleas off, and saved him; and the cat supported her through her own medical struggles for sixteen years. Finally, diabetes necessitated euthanasia for her pet, and my e-mail had reminded her of her lost friend and the tears streamed down her face.

Some were funny. "Jeff," one woman wrote, "are you married?"

When I showed that one to Lynda, she grinned and said, "Have her send a picture."

Many people wrote and told me about a family member who was sick and dying and needed help. Some were grasping at anything they could, trying desperately to save themselves or a person they loved deeply. People asked for Freedom to heal their loved ones.

I was apprehensive of people seeing Freedom as a

healer—even as I understood their pain and hope. I would write back and explain that although I believe that Freedom's love for me made it possible for her to help me fight my cancer, she cannot bend the laws of nature. If I had needed proof, I could only reflect that Kaye and Bob would still be with us today if she could have saved them.

Freedom is an ambassador for her species and a teacher at our educational programs. Her ability to travel is limited, and she needs privacy and insulation from too much exposure to strangers. I would never risk her health or safety through overexposure, not even to give hope to the sick and dying.

But when I read e-mails that say things so brutally honest, of pain intense and very real, like that from the single mom of two teenage girls who wrote me, "Thank you for sharing your gift of life and will. I was diagnosed right before my 42nd birthday, with metastatic colon cancer. . . . I have been through the gamuts of chemo and surgery. . . . I don't think enough of us stop in the moment like you did with Freedom to appreciate the gifts we are given," I almost wish that Freedom *could* heal the sick. I could only read a few e-mails like that at a time.

Many people wrote just to say thank you for sharing Freedom's and my story with the world. A woman who worked as a women's advocate for domestic violence and sexual assault wrote that after a difficult morning "my heart was so heavy and breaking that I really considered going home but I came back to my office and read your story and realized that God was talking to me and that he always shines the light when we need it the most." This was in the first few months and I was still trying to answer all the e-mails as they came, but I couldn't keep up with the deluge. I am touched that so many thought to write me and share their stories.

Some of these writers became important in my life. On March 20, 2008, I got this message:

> Dear Jeff,
> This is the kind of story that helps me connect with my self in ways that are not explainable. I am a dog musher and I work with dogs, and kids who have emotional issues. The dogs and the kids seem to create a healing environment together and it's a really wonderful thing. Thank you for your courage with Wambli Oyate [eagle nation].
>> All my respect to you and Freedom,
>> Stephanie Little Wolf

I wrote her back, right away, my curiosity piqued by her work and her use of the Lakota language. Her therapy program, medicinedogs, helps Alaskan native children in foster care, near where she lives in Fairbanks. We hit it off right away; I felt like I do when I run into an old friend I haven't seen in years.

Stephanie became a *kola*, a friend. I mentioned to her that I had been searching for Freedom's "real name or spirit name" since she was a young bird and it never came to me. I had asked Kaye once to see if a native name would come for Freedom. At this time I didn't know much about native traditions and naming. The right name just didn't come. I've learned since that a native name has to present itself; you can't go looking for it, but you know when it's right.

Stephanie is Lakota, Shoshone, and Yaqui. I had a hunch she might know what Freedom's real name was.

She wrote me back, "What about Dream Flyer?" And she told me how to say it in Lakota.

I was at home when I read Stephanie's e-mail at my computer before dinner. I looked at it again and again; I couldn't believe the simplicity and beauty of this name. It was perfect, and it was exactly who Freedom was. And only Stephanie could see it from the outside, beyond my searching.

The right name to call my friend wasn't the only gift of these e-mails. They connected me to many people—including the people who helped me write this book.

The e-mails came from the Americas both north and south, Europe, Africa, Asia, even from Americans in the field and on air bases in the war zones of Afghanistan and Iraq. Freedom has been on the covers of magazines in Australia, New Zealand, and South Africa and here in the USA. She has had offers to travel the United States and Europe. She's received electronic art from Tel Aviv, and an offer for a late-night TV appearance in Holland. Children in Japan interviewed us for their schoolwork. Kids in a Michigan elementary school wrote a song for her. Newspapers from Calcutta to London and major TV networks all have wanted this story—a story that is simply a tale of love and hope.

It was hard to comprehend the outpouring of emotion coming our way. How I found myself in this position was amazing to me. I had never expected anything back from that e-mail. I was just telling a friend a love story.

CHAPTER FOURTEEN

THE HEAVY SNOWS STARTED on December 14, 2008, and didn't let up. Four days later Freedom cut back noticeably on her eating—which could have been particularly dangerous for her in the cold weather. She completely quit eating on December 21, 2008. She'd walk away from her food and when she came back to it, even if she had wanted to eat, she couldn't—it would be frozen and inedible.

As the snow kept falling, Freedom talked to Wanbli Askata in eagle speak. There were *chak chaks*, chirps, and screams. Although it was common for the two eagles to converse, this time Freedom was much chattier. Askata, who's usually a glutton, stopped eating on the twenty-second, a day after Freedom had quit—I had a hunch that Freedom had told her to stop.

The weather was in the high teens for the day and single digits for the night. By the twenty-third—when Freedom hadn't eaten for two days—our clinic director, Leslie Henry, decided to have her pulled inside. "I think that's our only choice," she said to me. "We can't let anything happen to this bird." I agreed. We both knew that weather that cold can sap eagles' energy reserves and leave them vulnerable to potentially fatal dehydration.

When Freedom came inside, she already needed to be rehydrated. We did that by tube feeding her. She hadn't been tube fed since she was a baby and she didn't like it. We finished feeding her and put her back in the dowel cage, where she had started her life at Sarvey so many years before. She started eating again the next day.

Leslie called me at home on Christmas Eve and told me Freedom's flight had just collapsed. She said the sound of ripping metal and snapping four by fours sounded like a freight train tearing through the clinic. The sound had brought her and a number of others running outside. Freedom's flight was in ruins—if she had been inside, she would have been killed.

"Shit!" I said. I knew the rebuilding would be tough and that Freedom would have to stay inside for months, which both of us would hate. Then I realized I had so much to be thankful for. Freedom was alive. Flights can always be rebuilt.

Askata's flight next to Freedom's was still standing—but no one knew for how long. Askata was hunkered down in a corner. The staff scrambled to get her out of there before it came down. Askata didn't resist them. She was docile and afraid. Her flight collapsed the next day.

Askata and Freedom weren't the only eagles inside. We'd also had to bring in three eagles who'd been rehabbing in the large eagle flight.

By Christmas day we had lost the waterfowl flight, the large eagle flight, Freedom's and Askata's flights, and the otter pen. There was severe damage to the cat pen roof. The large mammal enclosure where we keep bears was in danger of a roof collapse, and many more flights were threatened. All the wild ones were safe, but the repairs were going to be extensive.

Christmas morning, I drove to Sarvey. The snow was still coming down hard. It was clear there was going to be much more work to do. Leslie was already at the center along with Kelly, Tammie, and Angie, a brand-new staff member. We had four or five feet of snow in spots. We couldn't get to some flights at all. The clinic roof lay under three feet of snow. I was worried. If it imploded, it would be a disaster. Freedom and Askata were in there along with the other rehab eagles and injured patients. The center was turning into a war zone.

Leslie and Angie shook snow off the golden eagle's flight.

Kelly and Tammie did daily rounds of the patients, then joined the battle against the snow on what was left of the flights. I got on the roof of the clinic with a broom and a shovel. Almost every shop broom and shovel we had had been broken by falling debris; the ones I had were all that were left.

We left the empty flights and enclosures alone to focus on knocking snow off the ones whose occupants we couldn't bring inside. It was an all-day job for the staff and volunteers who had been able to make it into Sarvey. Kelly couldn't go home—her car was buried under snow. She stayed with Tammie, who is the nighttime on-call medical person and lives in Kaye's old house. They couldn't get out for food, so some of the volunteers and staff who could get home brought in food for the next few days. The snow kept coming down, even as the weather reports said it wouldn't.

We had parked the ambulance at the bottom of the hill because of the snow. We didn't want it to be stuck at the top if it was needed for a rescue—which turned out to be a wasted effort since it couldn't go anywhere in the snow. Right next to the ambulance was a tree that Kaye had especially loved, an old ash tree that had grown to over twenty feet high and with so many branches fanning out it was almost as thick. It was a great shade tree. It shaded the edge of the dirt drive that led up the hill. With the weight of more snow than it had known in many years, the ash tree crashed down hard onto the ambulance, burying it and pulling the main power line down on top of it.

Another tree Kaye had especially cared for stood at the top of the hill, an old haggard apple tree that bore fruit every year and provided shade for the garden underneath. Kaye always had paid it special attention, cleaning up underneath it and making sure it was healthy and green. The weight of the snow took that tree down too.

Standing at the top of the drive in mounds of snow, I saw the ash tree on top of the ambulance, and the apple tree next to me on the ground—torn in half. I saw the wreckage of the eagle flights, and trees hanging on the last power line coming into Sarvey.

It hit me. This storm was a rebirth for Sarvey, and a test. Kaye was gone. Grandfather had taken her trees. Sarvey Wildlife Care Center, savior of lives and souls, belonged to those who were left—and to the wild ones. It was up to us humans to go forward.

The snow stopped at last. When we dug out the ambulance, we discovered it was unscathed. We hired Mark Olsson, who had built cages and flights for us in the past on a volunteer basis, to rebuild everything.

Our January monthly cleanup day stretched to two—and required a big dump truck to haul tons of debris to the dump. Then Mark began the rebuilding process. He drew up the plans for the new eagle flights and, after the center's approval, went to work immediately.

We knew the expenses would mount up in a hurry, so we would need more money than came in from our regular contributors. Kelly Pattison, our administrative director, spearheaded a fund-raiser. Her efforts brought in more than we could have hoped for. We were grateful. The rebuilds and repairs were going to be costly. They would take most of the year, but we had no choice. Everything would be overbuilt— to weather the most severe storms and last far into the future.

Freedom stayed in the dowel cage for three months while her new indestructible flight was being built. She was a little too docile, and I wondered if she felt she was being punished. I took her outside twice a week to keep her waterproofed and acclimatized and so she could breathe the sparkling winter air.

Volunteers let her out of the dowel cage a few times a day. That was her cue to run around and bug people cleaning cages. She also picked up a new habit of playing with apples. She'd hop around with the apple clutched in her talons then let it roll away and go get it. At the end of the game she'd bite a chunk or two off the apple and eat it.

Sarvey's new chapter began in March 2009 for me, when I took Freedom to her new home. It was much bigger, and swanky, with a hideaway house built inside that she could use to get away from everyone—and a brand-new pool.

Did someone say pool? She had been taking bird baths inside and hadn't had a real bath in three months. Now it was time for a bath with a vengeance. She normally prefers to bathe alone, but that wonderful day a proper bath could not be delayed. She sauntered over to the pool, hopped in, looked around, and dove underwater. She popped her head in and out and madly shook her feathers. It was pure joy for her—and for me to watch.

I know that as the snow continued to fall, Freedom had understood that her flight was a dangerous place. Her fast saved her life—it drew attention and made it obvious she would have to come inside. I know that she understood that Askata's flight had also become dangerous, and she told her friend to stop eating. I trust that knowledge.

I want to honor both her knowledge and my trust. It has now been eleven years since Freedom came into my life. I can't imagine living without her. I have seen grown men and women walk up to her and start sobbing. I have seen the fierce protector in her watching over me. I have seen the comedian eagle play jokes on people. I have watched her take over crowds with antics that let humans know, "I understand."

As I began writing this book, I was thinking so often about what Freedom has brought into my life—and how I could return the gift. The answer came from an unexpected source. One morning, as I was getting out of my truck at work, a big SUV drove by with a bumper sticker that said CHANGE COURSE. The irony was all too obvious—though maybe not in the way the occupants of the SUV meant it. I knew how I wanted to end my book.

We all take from this world we live in—and are a part of. In our race to have everything, we lose a little every day. There is a whole world we pay no attention to and it is ours to marvel at. Day by day the wild ones are losing ground—and their lives—to human development and all that goes with it. The wild ones communicate with us constantly, but we have forgotten how to see them, much less understand what they are telling us. Their message is critical, as is their presence.

It is time for us to change course before it is too late. Without the wild in our lives, we will have lost our species' connection to the circle of life. I remember how Kaye would close each educational presentation with a beautiful Lakota prayer. She would place her hand over her heart, palm up, make a sweeping circle back to her heart, and say "Mitakuye oyasin." That means "I will live with all my relations." And "All my relations" means the ones that fly, the creepy crawlies, our rooted friends, the four-legged, the two-legged, the ones that swim, the very earth we live on.

I will live with all my relations—Mitakuye oyasin. Kaye believed that humans have strayed a long distance from this elegant vow. So she would have all the kids speak the vow. They understood—they could grasp that they are the future. They could learn that if our wild world is to survive, they will have to bear the torch.

I was finishing this book on the Fourth of July—exactly one year after Kaye's death. Sadness and hope permeated my being. I stopped writing and closed the computer. It was time to go to Sarvey to pick Freedom up and bring her home— because the fireworks, visible from her flight, still scare her so badly.

I hopped into my truck and headed out for the center. As I was driving down Highway 2 the sun was just beginning to drop to the horizon. The sky was filled with thin hazy clouds turning yellow. As I drove westbound I saw a band of clouds stretched across the sky that looked as though it had minia-ture tornados hanging from it. I thought of my drive the year before—on the day Kaye died.

When I arrived at Sarvey, Freedom was animated. She greeted me with her head held high and a chorus of *chak chak, chak chak*. We talked together for a minute and then I went into the clinic. Angie, who had gotten her baptism by fire— or snow—back in December as a new employee, was working on a pigeon in the med room. "Go look in large green five," she said. "She came in a few hours ago."

I walked over and looked in. There was a very young ea-gle about three months old. She was standing and her wings were fine. Too young to fly, the baby eagle had fallen from her nest. She was emaciated and lice ridden. I looked at her and was transported back eleven years to the day Freedom had first arrived.

I bent down and whispered, "Hey, how are you my wild one?" Her baby eyes stared at me. I could see the fear in them.

I backed away. Even though this kid knew she was an eagle, it was best to keep human contact to a minimum to lessen her stress. She should not be habituated to humans because she would probably fly with her brothers and sisters someday soon.

I went to Freedom's flight, took her out, and walked to the front of Kaye's old house. I looked at the window where Kaye had been sitting when she died. My mind was full—Freedom and me, a new baby inside, an elder who left to make room for the new ones. I asked Kaye if she heard my thoughts and if she saw how we were carrying on her legacy.

The light was fading fast. It was time to go. I opened the carrier door in the back of my truck. Freedom jumped inside, rarin' to go. I backed down the drive and we headed out. As always, I was conscious that I carried precious cargo. The fireworks had started, and the sky was blazing with explosives. I wished all my wild friends safety. This human holiday is terrifying for them.

I turned onto Highway 2, the highway of death. I thought of my ride on this very road one year ago. I rounded the bend where my deer brother was killed. Had that been a dream or reality? One year was merging into the next. My thoughts were jumbled. The question repeated again and again. *It was real*, I thought to myself. I knew I had seen the blood and fur the next morning. Even though the distance of a year made it feel like a dream, I knew my experience had been real. I drove on.

It was full night when I pulled up to the house and parked. I got Freedom squared away on her perch in the living room. Lynda came in from feeding our three cats and got comfortable on the couch across from me. The cats followed her in, took one look at Freedom, and beat feet upstairs. Freedom sat Queen over it all. Life was as it should be, surrounded by mystery and love

Mitakuye oyasin.

NOTE TO THE READER

Wild and tame are not the same." This is what Sarvey's educators tell kids at every school program. And we ask them what it means. The answer we want is, "You can never make a wild animal tame." A wild animal has thousands upon thousands of years of instinct, and a person thinking he can bottle-feed a wild cougar or bear and tame it is delusional. The animal's wildness will always be there just under the surface. Wild ones like Freedom and Sasha the cougar *allow* us to interact with them; it's their choice. They are not tame and never will be. It is cruel to condemn a healthy wild one to a life in captivity, and dangerous too. Strict protocols govern our work at Sarvey. I have had great joy from Freedom, from Sasha, from Angelica the bear, but if I had a time machine and the omnipotent ability to change their lot, I would trade it all to have never met any of them, to let them live their lives as wild ones should—unbroken, unconfined.

Wildlife rehab takes commitment and courage. I hope some of my readers may be inspired to join us in the work. You will have your heart broken and you will witness the power of every animal's instinct to survive. Your heart will explode with joy at seeing a wild one released. We can never forget that watching an eagle soar into the heavens or a squirrel scamper up a tree is why you do this. Freedom and I are fighting for the heavens, to ensure that the skies will be unpolluted and safe for eagles and that the trees will still be there. The wild ones ask for nothing from us—only to live their lives in this ever-shrinking world of clear-cuts and paved-over lands that used to be their home.

We at Sarvey love our work—but we all wish we could be out of business. If we weren't needed, this world would be a much better place. But we are needed—more and more every year. Every wildlife center in this country is struggling for money and volunteers.

If you can, help support a wildlife center near you. Learn about the wildlife in your area and what to look for. Learn how *not* to rescue babies whose mothers are only out of sight, like Angelica the bear's, or the many fawns that come to Sarvey every year. We all take from this world, and that's part of living in it; it just seems fair to give a little back.

Wild ones communicate on a level we stray away from more each day. They "speak" to us all the time; they have emotions and show them. If you take the time to watch and listen, you will learn and understand.

To find out more about Freedom and where she lives, go to www.sarveywildlife.org.

I hope you enjoyed this book as much as I enjoyed writing it.

LIVE WILD AND FREE,

JEFF GUIDRY

ACKNOWLEDGMENTS

WHEN MY E-MAIL WENT out to Gayle it was really as an afterthought, and to a friend. That it touched raw emotions of love and hope and made people feel good surprised me. I know now it shouldn't have. The journey Freedom and I are on is one of mutual love, learning, and respect. I am humbled to know this magnificent bird, and honored for her to accept and bond with me in her world. I never had any intention of writing this book, but it became clear early on this was a story many folks would enjoy and many asked me for this book. Thank you to everyone who wrote, forwarded, or just enjoyed our saga. This book is for you. Without your honest response, there probably would be no book. And without the people listed below, there wouldn't either:

All Sarvey volunteers, past, present, and future, for the

love you give freely to the wild ones and for all the hard work and blood that goes with it. Most of all, I thank Kaye and Bob, who live forever in this place of compassion, heartache, and joy.

To Jennifer Pooley, editor magnifique! I could not have asked for better. She ran this show with a deep understanding of how this book should flow, and was exceptional in guiding and helping me through the whole process. Jen had faith in me and in this story, and she made me believe and made me better than I thought I could be. For her compassion and vision, thank you; for her brilliant work editing this book, no quarter given, I am forever grateful.

Jean Marie Kelly, for loving this book so much you had to have it. To Shawn Nicholls, Brianne Halverson, Lynn Grady, Liate Stehlik, Nancy Tan, Richard Aquan, and Joseph Papa for all their inspiration and dedication to this project.

To Kate Epstein, agent extraordinaire. How insanely lucky was I to get Kate as my agent? I won the lottery. From the wee hours of the morning into the late nights and weekends, Kate gave completely of herself. She helped me do multiple edits and rewrites, time after time after time. The blood-letting never stopped, but it was all for the book. This book would not have happened without her. Kate kept me sane against self-doubt, and she made me double over with laughter when I most needed it. Not only is she my agent, but she is my friend. I will always be in her debt.

To Liz, my friend and teacher. What can I say? I can never repay you for what you have shown me. You have laid upon me a fire and a gift that I will always have. I had lost my way and you opened that creative door again. This book would never have been done without your coyote coming through and howling every time I needed inspiration or a laugh. You helped me find my voice in this writing and you trusted me. I feel extremely lucky to call you friend and know we always will be.

Dr. Andrew Jacobs, for giving me a second chance at life.

Annie Marie Musselman, for your uncommon vision and stunning photography that grace the cover and pages of this book. You have the gift of being able to bring out the vulnerable side of humans and wild ones in your art. Your work stands alone.

Judy Buzeck, for teaching me how to catch eagles . . . and how not to; for the cookies and for being a great friend all these years.

My little Angelica, I so hope you have greeted many a winter and had cubs of your own. I can imagine you still roaming the mountains and valleys. You were an absolute joy to experience and you will always be in my heart.

Leslie Henry, the director of Sarvey Wildlife Care Center, for pulling Freedom in before the crash and for your devotion to the wild ones. I can't count the number of times I called Les and asked questions to get the medical facts in the manuscript right. She always came through, no matter how busy she was.

To Kelly Pattison, my partner in crime at Sarvey. That place could not function without her. She brings an intensity and zeal to make Sarvey better all the time for the patients and the volunteers.

Sue McGowan, for all the years of looking out for Freedom. And for your undying love of squirrels. The shadow tails couldn't have a better friend.

Ken Lubas, for all your beautiful art that you have donated to help this place grow.

Kestrel Skyhawk, for all your years of selfless work at Sarvey. And for being the "bad guy" coping Freedom.

For their dedication to the wild ones, help with this book, and their friendship, in no particular order: Myron Phillips, Sharon and Ralph Akres, Pegi Smith, Alycia Leonard, Sandy Dahl, Amy Stolzenbach (aka Pimperella), Jessica Lazaris, Tammie Rohr, Angie Koellner, David Farage, Douglas Farage, Michaela Adams, Christine Szypulski, Sally Maughan, Billy Stapleton, Roni Stapleton, Bob Kamoske, Ellen Kamoske, Kate Byrnes, Jorge Harada, Liz Herrin, Jamie Farage, Jimmy Thomas, Thom Beebe and P.K., Tami Tate Hall (USF and WS), Martha Norwalk, the Stillaguamish tribe, Ed White (King 5 news, Seattle), Peggy Jewell, Dorian Tremaine, Mark Olsson, Cindy Palmiter, Caryn Rea, Gary Bullock and everyone at BT and LC in Anchorage, Carol Asvetes, Summer

Matthes, Barb Ogaard, Ron White, Amber Chenoweth, Jeff
Giesen, Pamela Bishop-Sholty, Will and Jean Hobbs, Chad
Hanson, Denise St. John, Kaye Hickman, Dianne Johnson,
Isaara Willenskomer, Jeff Butts, Eric Walton, Kevin King,
Becky Curtis, James Harry Smith, Lisa Ramondi and Koko,
Loraine and Garry Robbins, Paula Hegedus, Norbert Trahan,
Terry Trahan, Jeanne Pascal and Dallas, Jim Fannon, Natalie
Paige, Jeff Medicine Bear, Michael Cotta, Steve Kelly, Helen
Bishop, Bernadette McCrea, Corrina Gillette, Cark Knud-
sen, Robert Young, Paul Balle, Lorna Doone, Matt Blair, Ron
Bowman, Elizabeth Kamaka, Mike Robertson, Angeles Pena,
Billy Kennelly, Chris Quarrella, Greg Black, Libby Mills,
Paul Black, Miceal Wenke, Andy Aldrich, Rocky Spencer,
Trish Shallest, Jim and Ceal Kight, Randy Fader, Boyce and
Ann Trahan, Angela, Steve, and Taylor Deshotels, Little Bill
Engelhart, Jodi Head, Damon Logan, Will Miller, Muraco,
Jet Pagano, Bahaa Sadak, Michael Green, Cindy Lahey, Terry
Weaver, Dick Everist, Randy Ross, Donny and Vicki Evola,
Greg Keplinger, John Loftus, Reed Hutchinson, Sheryll
Gregory, Kirk Schroder, Brenda Kolb, Lori Lloyd, Arthur
Lee, Theo, and Denali.

Stephanie Little Wolf, for the gift of friendship and Freedom's spirit name, and for letting me reproduce her thought-
ful e-mail in this book. *Pikicila.*

Gayle Hoenig, for sending my e-mail out and starting
this whole crazy ride.

Nancy Bailey, for the work you did, and to Clifford.

Melanie Graham, for introducing to me to Sarvey and Kaye. Without you doing that, none of this would have happened. It's funny how everything is connected in some way. Look close enough, you'll see the thread.

Jill Black, my sister and oldest and dearest friend, you are the best! I can't imagine life without you.

Mom and Dad. I wish more than anything my dad was here to meet this eagle and to see our latest crazy ride. My mom, for always putting up with my menagerie (even when the snakes got loose in the house), and to both of them for instilling in their children respect for all living things and a love of wildlife.

To Grandfather, for sending the Dream Flyer on her mission.

For my best friend and love of my life, Lynda Robertson. Without you I wouldn't be alive today. You were there by my side fighting with me and supporting me in those evil days of the cancer. You kept my spirit up; you were my strength. I can't imagine how you felt, but I know how lucky I am to have you. When I began writing this book we didn't see each other much. I would come home from work, head upstairs till midnight or later and do it again the next day, six and seven days a week, for months on end. You were the "book widow." I've missed you. Thanks for sharing your life with me and putting up with all I do.